VOLUME FOUR TWENTY SEVEN

METHODS IN
ENZYMOLOGY

MicroRNA Methods

METHODS IN ENZYMOLOGY

Editors-in-Chief

JOHN N. ABELSON AND MELVIN I. SIMON

Division of Biology
California Institute of Technology
Pasadena, California

Founding Editors

SIDNEY P. COLOWICK AND NATHAN O. KAPLAN

Methods in
ENZYMOLOGY
MicroRNA Methods

EDITED BY

JOHN J. ROSSI
Graduate School of Biological Sciences
Division of Molecular Biology
Beckman Research Institute of the City of Hope
Duarte, California

GREGORY J. HANNON
Cold Spring Laboratory
Cold Spring Harbor, New York
Howard Hughes Medical Institute
Chevy Chase, Maryland

AMSTERDAM • BOSTON • HEIDELBERG • LONDON
NEW YORK • OXFORD • PARIS • SAN DIEGO
SAN FRANCISCO • SINGAPORE • SYDNEY • TOKYO
Academic Press is an imprint of Elsevier

ELSEVIER

Academic Press is an imprint of Elsevier
525 B Street, Suite 1900, San Diego, California 92101-4495, USA
84 Theobald's Road, London WC1X 8RR, UK

This book is printed on acid-free paper. ∞

For information on all Elsevier Academic Press publications
visit our Web site at www.books.elsevier.com

ISBN: 978-0-12-373917-9

PRINTED IN THE UNITED STATES OF AMERICA
07 08 09 10 9 8 7 6 5 4 3 2 1

Contents

4. Computational Methods for MicroRNA Target Prediction   65

Yuka Watanabe, Masaru Tomita, and Akio Kanai

Section II. MicroRNA Expression, Maturation, and Functional Analysis   87

5. *In Vitro* and *In Vivo* Assays for the Activity of Drosha Complex   89

Yoontae Lee and V. Narry Kim

6. Microarray Analysis of miRNA Gene Expression   107

J. Michael Thomson, Joel S. Parker, and Scott M. Hammond

13. Protocols for Expression and Functional Analysis
of Viral MicroRNAs   229

Eva Gottwein and Bryan R. Cullen

Contributors

George Adrian Calin
Department of Molecular Virology, Immunology and Medical Genetics and Comprehensive Cancer Center, The Ohio State University, Columbus, Ohio

Chang-Zheng Chen
Department of Microbiology and Immunology, Baxter Laboratory of Genetic Pharmacology, Stanford University School of Medicine, Stanford, California

Xuemei Chen
Department of Botany and Plant Sciences and Institute of Integrative Genome Biology, University of California—Riverside, Riverside, California

Carlo Maria Croce
Department of Molecular Virology, Immunology and Medical Genetics and Comprehensive Cancer Center, The Ohio State University, Columbus, Ohio

Bryan R. Cullen
Center for Virology and Department of Molecular Genetics and Microbiology, Duke University Medical Center, Durham, North Carolina

Eva Gottwein
Center for Virology and Department of Molecular Genetics and Microbiology, Duke University Medical Center, Durham, North Carolina

Adam Grundhoff
Heinrich-Pette-Institute for Experimental Virology and Immunology, Hamburg, Germany

Scott M. Hammond
Department of Cell and Developmental Biology, University of North Carolina, Chapel Hill, North Carolina

Hiroshi Imai
Laboratory of Reproductive Biology, Department of Agriculture, Kyoto University, Kyoto, Japan

Hailing Jin
Department of Plant Pathology, Center for Plant Cell Biology and Institute for Integrative Genome Biology, University of California—Riverside, Riverside, California

Akio Kanai
Institute for Advanced Biosciences, Keio University, Tsuruoka, and Systems Biology Program, Graduate School of Media and Governance, and Faculty of Environment and Information Studies, Keio University, Fujisawa, Japan

Surekha Katiyar-Agarwal
Department of Plant Pathology, Center for Plant Cell Biology and Institute for Integrative Genome Biology, University of California—Riverside, Riverside, California

V. Narry Kim
School of Biological Sciences, Seoul National University, Seoul, Korea

Saulius Klimašauskas
Laboratory of Biological DNA Modification, Institute of Biotechnology, Vilnius, Lithuania

Yoontae Lee
School of Biological Sciences, Seoul National University, Seoul, Korea

Haitang Li
Division of Molecular Biology, Beckman Research Institute of the City of Hope, Duarte, California

Tin Ky Mao
Department of Microbiology and Immunology, Baxter Laboratory of Genetic Pharmacology, Stanford University School of Medicine, Stanford, California

Naojiro Minami
Laboratory of Reproductive Biology, Department of Agriculture, Kyoto University, Kyoto, Japan

Joel S. Parker
Constella Group, Durham, North California

Sébastien Pfeffer
Institut de Biologie Moléculaire des Plantes, CNRS, Strasbourg cedex, France

John J. Rossi
Graduate School of Biological Sciences, and Division of Molecular Biology, Beckman Research Institute of the City of Hope, Duarte, California

Pål Sætrom
Interagon AS, Laboratoriesenteret, and Department of Computer and Information Science, Norwegian University of Science and Technology, Trondheim, Norway, and Division of Molecular Biology, Beckman Research Institute of the City of Hope, Duarte, California

Hiroyuki Sasaki
Division of Human Genetics, Department of Integrated Genetics, National Institute of Genetics, Research Organization of Information and Systems, and Department of Genetics, School of Life Science, The Graduate University for Advanced Studies (SOKENDAI), Mishima, Japan

Ola Snøve, Jr.
Interagon AS, Laboratoriesenteret, and Department of Cancer Research and Molecular Medicine, Norwegian University of Science and Technology, Trondheim,

Norway, and Division of Molecular Biology, Beckman Research Institute of the City of Hope, Duarte, California

Christopher S. Sullivan
Department of Molecular Genetics and Microbiology, University of Texas at Austin, Austin, Texas

Guihua Sun
Graduate School of Biological Sciences, and Division of Molecular Biology, Beckman Research Institute of the City of Hope, Duarte, California

J. Michael Thomson
Department of Cell and Developmental Biology, University of North Carolina, Chapel Hill, North Carolina

Masaru Tomita
Institute for Advanced Biosciences, Keio University, Tsuruoka, and Systems Biology Program, Graduate School of Media and Governance, and Faculty of Environment and Information Studies, Keio University, Fujisawa, Japan

Yasushi Totoki
Genome Annotation and Comparative Analysis Team, Computational and Experimental Systems Biology Group, RIKEN Genomic Sciences Center, Tsurumi-ku, Yokohama, Kanagawa, Japan

Giedrius Vilkaitis
Laboratory of Biological DNA Modification, Institute of Biotechnology, Vilnius, Lithuania

Toshiaki Watanabe
Division of Human Genetics, Department of Integrated Genetics, National Institute of Genetics, Research Organization of Information and Systems, and Department of Genetics, School of Life Science, The Graduate University for Advanced Studies (SOKENDAI), Mishima, Japan

Yuka Watanabe
Institute for Advanced Biosciences, Keio University, Tsuruoka, and Systems Biology Program, Graduate School of Media and Governance, Keio University, Fujisawa, Japan

Zhiyong Yang
Department of Botany and Plant Sciences and Institute of Integrative Genome Biology, University of California—Riverside, Riverside, California

Bin Yu
Department of Botany and Plant Sciences and Institute of Integrative Genome Biology, University of California—Riverside, Riverside, California

METHODS IN ENZYMOLOGY

IDENTIFYING MicroRNAs AND THEIR TARGETS

CHAPTER ONE

IDENTIFICATION OF VIRAL MICRORNAS

Christopher S. Sullivan* *and* Adam Grundhoff[†]

Contents

Abstract

Given the important function of microRNAs (miRNAs) in the control of gene expression, it is not surprising to find that some viruses encode their own miRNAs. Although the function of the overwhelming majority of these miRNAs remains unknown, at least some of them are expected to play crucial roles in the viral life cycle, and hence there is great interest in identifying novel viral miRNAs. The majority of currently known viral (and host) miRNAs have been identified by cloning of small RNAs, but, due to their small size, viral genomes are especially amenable to alternative methods based on the computational prediction of miRNA candidates. Here, we provide a detailed protocol on how to use computational prediction methods in conjunction with high-throughput microarray analysis to detect miRNAs in viral genomes.

1. INTRODUCTION

With the realization that miRNAs play a profound role in regulating the gene expression of most multicellular eukaryotes comes the understanding that some viruses that infect these hosts encode their own miRNAs. In 2004,

* Department of Molecular Genetics and Microbiology, University of Texas at Austin, Austin, Texas
† Heinrich-Pette-Institute for Experimental Virology and Immunology, Hamburg, Germany

Methods in Enzymology, Volume 427
ISSN 0076-6879, DOI: 10.1016/S0076-6879(07)27001-6

Pfeffer and colleagues were the first to report that viruses encode such molecules when they cloned miRNAs of viral origin from a B-cell line latently infected with Epstein Barr virus (EBV) (Pfeffer et al., 2004). Since then, over 80 virally derived miRNAs have been discovered, mostly in viruses that are members of the herpesvirus and polyomavirus families (Table 1.1). Both families contain viruses with deoxyribonucleic acid (DNA) genomes that can establish lifelong, persistent infections in their hosts. In addition, there is a report that a single strain of human immunodeficiency virus (HIV), a member of the retroviral family, encodes an miRNA (Omoto et al., 2004). (However, whether the processing of this small RNA is dependent on the miRNA machinery, how prevalent this small RNA is in other HIV strains, and whether this small RNA is relevant to the life cycle of the virus awaits further studies.)

As with their host counterparts, the functions of the vast majority of viral miRNAs are unknown. Clearly defined functions of three viral miRNAs have been established (Cullen, 2006; Pfeffer and Voinnet, 2006; Samols and Renne, 2006; Sullivan and Ganem, 2005). SV40 expresses a pre-miRNA late during infection that is processed into an miRNA and an miRNA★, both of which actively direct the cleavage of early gene products (Sullivan et al., 2005). This causes reduced early protein levels at late time points of infections, resulting in less susceptibility of infected cells to T-cell–mediated lysis in in vitro assays. Additionally, EBV encodes for miR-BART-2 that can direct the cleavage of the transcript encoding the BALF5 DNA polymerase (Pfeffer et al., 2004), which may suggest a common theme of viral utilization of miRNAs for autoregulation of gene expression. However, Fraser and colleagues have reported that HSV-1 encodes a miRNA expressed during latent infection that negatively regulates certain cellular transcripts involved in a proapoptotic response (Gupta et al., 2006). This suggests that at least some viral miRNAs will have a profound effect on the viral life cycle by targeting host transcripts.

The majority of known miRNAs (including those encoded by viruses) have been identified via cloning. In this procedure, small RNAs are size-fractionated, and successive rounds of single-stranded ligation incorporate flanking regions on the RNA to facilitate reverse transcription, PCR-mediated amplification, and cloning. The methodology for cloning viral small RNAs is identical to that for host-derived RNAs, and, therefore, we direct readers interested in cloning viral miRNAs to several established and well-written protocols (Pfeffer et al., 2003; http://web.wi.mit.edu/bartel/pub/protocols/miRNAcloning.pdf). However, the relatively small size of viral genomes makes them particularly amenable to computational-based miRNA discovery methods. Indeed, a sizable fraction (~15%) of virally derived miRNAs were identified solely by noncloning computational methods. Here we present the methodology behind our experimental strategy of using computational prediction combined with microarray detection to identify virally encoded miRNAs.

Table 1.1 Viruses known to encode microRNAs

Virus	Family	Number of known miRNAs	Method of identification	Reference(s)
Epstein Barr virus	Herpesviridae	23	17 by cloning, 6 by a combination of computational prediction and microarray analysis	Cai et al., 2006; Grundhoff et al., 2006; Pfeffer et al., 2004
Kaposi's sarcoma herpesvirus	Herpesviridae	12	11 by cloning, 1 by computational prediction and microarray analysis	Cai et al., 2005; Grundhoff et al., 2006; Pfeffer et al., 2005; Samols et al., 2005
Human cytomegalovirus	Herpesviridae	11	9 by cloning, 2 by computational prediction	Dunn et al., 2005; Grey et al., 2005; Pfeffer et al., 2005
Murine herpesvirus 68	Herpesviridae	9	All by cloning	Pfeffer et al., 2005
Rhesus lymphocryptovirus	Herpesviridae	16	All by cloning	Cai et al., 2006
Herpes simplex virus 1	Herpesviridae	2	1 by computational prediction, 1 by a combination of computational prediction, cloning, and colony hybridization	Cui et al., 2006; Gupta et al., 2006

(continued)

Table 1.1 (continued)

Virus	Family	Number of known miRNAs	Method of identification	Reference(s)
Marek's disease virus	Herpesviridae	8	All by cloning and deep sequencing	Burnside *et al.*, 2006
Simian virus 40	Polyomaviridae	1	Computational prediction	Sullivan *et al.*, 2005
SA12	Polyomaviridae	1	Computational prediction	Cantalupo *et al.*, 2005; Sullivan *et al.*, 2005
Murine polyomavirus	Polyomaviridae	1	Computational prediction	Sullivan *et al.*, unpublished
Human immunodeficiency virus	Retroviridae	1?	Cloning	Omoto and Fujii, 2005; Omoto *et al.*, 2004

2. COMPUTATIONAL PREDICTION OF VIRAL MIRNA CANDIDATES

2.1. Principles of computational miRNA identification

Soon after the initial discovery of miRNAs, it was realized that bioinformatic methods hold great potential to aid in the identification of novel miRNAs, and a large number of miRNA prediction algorithms have been developed since then. By definition, miRNAs are small, noncoding RNAs generated by the RNAse III-like enzymes Drosha and Dicer (Ambros *et al.*, 2003). They are initially transcribed as part of much longer precursor transcripts (pri-miRNAs), and structural analysis has shown the region encoding the mature miRNA folds into a hairpin structure (the pre-miRNA) of approximately 60 to 80 nucleotides. Neither pre-miRNAs nor pri-miRNAs share any recognizable sequence motifs, and it is thus thought that the hairpin structures themselves serve as the primary signal that recruits Drosha and thereby initiates miRNA maturation. Accordingly, all miRNA-prediction algorithms use secondary structure analysis to identify potential pre-miRNA fold-back structures. Such predictions, however, are greatly complicated by the fact that hairpins are extremely abundant structures within any sequence. The problem, then, is to define features that distinguish pre-miRNAs from random hairpins or other structured RNA elements. According to a number of miRNA-defining criteria set forth by a consortium of researchers in 2003 (Ambros *et al.*, 2003), miRNA precursor hairpins "should not contain large internal loops or bulges, particularly not large asymmetric bulges." Although fulfillment of these minimal criteria is considered sufficient to prove that a small RNA identified via cloning indeed represents a miRNA (as opposed to a small interfering RNA[siRNA]), they are certainly not stringent enough to serve as a basis for computational prediction methods. Therefore, most miRNA prediction programs employ scoring algorithms based on the statistical comparison of a reference set of known pre-miRNA hairpins versus a control collection of unrelated hairpins. Minimally, such comparisons include consideration of bulge size and symmetry, but many algorithms also analyze additional features (e.g., sequence composition, position of internal bulges, length of helices, hairpin symmetry). The number of analyzed features as well as how they are weighted in the calculation of a final score differs considerably between the various prediction methods, which can result in little overlap between the predictions obtained with different programs. Although the predictions of some programs might be more accurate than others, it remains a fact that none of the currently available computational algorithms is able to reliably identify miRNAs based on structural analysis alone. Therefore, most miRNA prediction programs employ additional filtering methods to eliminate false-positive candidates.

By far the most frequently used filter is evolutionary conservation: because the overwhelming majority of miRNAs are located in nonprotein coding regions, pre-miRNAs often register as isolated islands of high-sequence conservation against the background of nonconserved DNA. Although such filters are very powerful, they have the obvious disadvantage of being unable to identify nonconserved miRNAs. Depending on the scope of the particular analysis, this trade-off between gain in specificity at the cost of sensitivity is often an acceptable one, especially given the majority of hitherto known animal and plant miRNAs that indeed appear to be conserved. However, for several reasons, such approaches are less suited for viral miRNAs. First, for many viruses, only very distant relatives are known, and thus suitable sequence information to conduct a comparative genome analysis is often not available. Second, viruses represent organisms that are highly adapted to their hosts, and even closely related viruses have frequently adopted different strategies to infect their hosts, by targeting different cell types or tissues, for example. As a result, many viruses harbor genes or genomic segments that are not or only very poorly conserved (e.g., the latency genes of many herpesviruses). It is reasonable to assume this is also true for viral miRNAs, and focussing only on evolutionarily conserved sequences thus might miss many miRNAs. These concerns notwithstanding, evolutionarily conserved viral miRNAs certainly exist: for example, EBV and rhesus lymphocryptovirus (RLV) share seven miRNAs (Cai *et al.*, 2006). However, the remaining EBV and lymphocryptovirus (LCV) miRNAs appear to be unique, and although a study that used phylogenetic comparison identified five miRNAs conserved between chimpanzee and human cytomegalovirus (CMV), it missed several others that are apparently not conserved (Grey *et al.*, 2005). Lastly, viruses often use their coding capacity to the maximum, and it is not unusual to find overlapping open reading frames (ORFs) on both strands of a given region or even in different reading frames on the same strand. Thus, it can be expected that at least some viral miRNAs overlap with ORFs (as indeed some of the known ones do), and because coding regions are masked in searches for evolutionarily conserved miRNAs, such candidates would be missed.

For these reasons, it is preferable to use *ab initio* prediction methods to identify viral miRNAs. However, such methods are naturally less accurate in their predictions. Two strategies are generally used to cope with this problem: the first is to increase the stringency of the algorithm scoring the hairpin structure. This will result in a decreased rate of false-positive predictions at the expense of false negatives such that bona fide miRNAs might be missed. The alternative is to perform a low stringency primary prediction, thereby allowing relatively large numbers of false positives followed by a high-throughput experimental screen to eliminate the false-positive candidates. The latter method is especially attractive for the identification of viral miRNAs because the small genome sizes of viruses will, even with minimal filtering, generate only a limited number of predictions that can be fitted easily on today's high-density microarrays. With these considerations in mind, we have previously developed a program called VMir to predict viral miRNAs.

2.2. Predicting viral miRNA candidates with VMir

VMir represents a low stringency prediction method for the identification of miRNAs in viral genomes (or other sequences up to approximately 2 Mb in size), which contains an easy-to-use graphical user interface. The program is described in detail by Grundhoff *et al.* (2006). Briefly, VMir performs an analysis by sliding a 500-nt window in steps of 10 nt over the sequences of interest. Within each window, the program performs a structure prediction by minimal free energy folding (using the RNAfold [Hofacker *et al.*, 1994] algorithm) and identifies individual hairpins above a given size cutoff (by default 45 nt). These hairpins are then scored based on a statistical comparison to a reference set of known pre-miRNA hairpins (see Grundhoff *et al.*, 2006 for details). The scoring algorithm of the VMir program was designed to overpredict rather then underpredict potential pre-miRNA candidates in viral genomes. As indicated previously, the ratio-nale behind this design was that the relative small size of viral genomes would permit such nonstringent filtering, thereby minimizing the risk of false-negative predictions. The VMir program nevertheless incorporates several user-adjustable quality filters, which can be used to reduce the complexity of the prediction. As an example, Fig. 1.1A shows the primary output from an analysis of SV40, a relatively small virus with a genome of approximately 5 kbp in which a total of 109 hairpins are detected. The results can reasonably be filtered by setting a score cutoff of 115 because 95% of the known pre-miRNAs in the training set reach or exceed this value. Under these condi-tions, only 16 hairpins would remain, a number that can be easily tested by Northern blotting. Another method to filter the results is independent of the scoring algorithm but, rather, is based on the robustness with which the structures fold in different sequence contexts. VMir analyzes windows that are significantly larger than the expected size of a pre-miRNA hairpin (60–80 nt). Furthermore, each sliding window overlaps with the previous one by 490 nt and thus a hairpin may fold in multiple windows (e.g., provided it is not located at the extreme ends of the analyzed genome, a hairpin of 80 nt will have the opportunity to fold in up to 43 windows). Because significantly stable structures are expected to fold in the majority of sequence contexts represented by the different windows, the number of windows in which a given hairpin is detected (referred to as the window count) can be used as a quality criterion. We consider this method preferable to a filter based on minimal free folding energy values (MFEs) alone because, depending on their nucleotide composition and length, even random sequences can have quite low, and therefore seemingly significant, MFEs. With a window size of 500 nt, we typically use window count cutoffs between 10 (least stringent) to 35 (most stringent). Under the most stringent conditions, only two

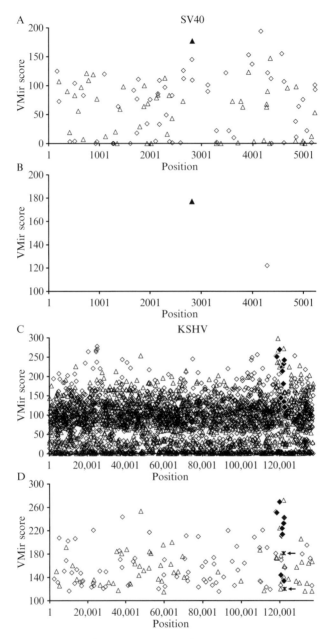

Figure 1.1   VMir analysis of the SV40 (A, B) and KSHV (C, D) genomes. Hairpins are plotted according to genomic location and VMir score. A and C represent the unfiltered output from the VMir prediction, whereas the results were subjected to stringent filtering (score and window count cutoffs of 115 and 35, respectively) to produce the plots shown in (B) and (D). Triangles and diamonds represent hairpins in direct or reverse

candidates (one of them the bona fide SV40-miR-S1 [Sullivan *et al.*, 2005]) remain in the analysis of the SV40 genome (Fig. 1.1B, filled triangle). Of course, larger genomes will produce significantly more candidates, and even stringent filtering might be insufficient to reduce the number of candidates such that each of them can be verified individually by Northern blotting. Figure 1.2C shows the output from a VMir analysis of the KSHV long unique region (LUR) which is ~140 kbp in size. 3046 hairpins register in the primary analysis, and 146 of these survive the stringent filtering method (Fig. 1.2D). Out of the 12 pre-miRNAs known to be encoded by KSHV (Cai *et al.*, 2005; Grundhoff *et al.*, 2006; Pfeffer *et al.*, 2005; Samols *et al.*, 2005), 10 are contained in the filtered prediction (shown as black diamonds in Fig. 1.2D), and 8 of these map to the top 20 scoring hairpins. Although the latter would have been surely identified in an experimental verification attempt based on the top scoring candidates alone, two pre-miRNAs with lower scores of 134 and 144 would likely have been missed. Likewise, two pre-miRNA hairpins, which fold in less than 35 windows and thus were eliminated from the analysis would also have evaded detection (their positions are indicated by asterisks marked with arrows in Fig. 1.1D). Thus, if Northern blotting would be used as the sole confirmatory test, bona fide miRNAs might be missed due to the stringent filtering necessary to produce manageable numbers of candidates. In contrast, microarray analysis provides the opportunity to investigate much larger numbers of miRNA candidates. Depending on the size of the viral genome and the capacity of the microarray format used, it is possible to screen the totality of all hairpin structures within a viral genome (which VMir will report if no filter is used). Thus, such an analysis can serve as a confirmation method that has the capacity to detect all miRNAs, regardless of the score awarded by the prediction algorithm. In the following section, we will provide a detailed protocol on how to design and analyze such microarrays.

3. ARRAY CONFIRMATION OF VIRAL miRNA CANDIDATES

The procedure described here can be divided into three steps: First, custom arrays are designed with aid of the VMir software. These arrays are then hybridized to small RNAs isolated from infected cells, and the

orientation, respectively. Confirmed pre-miRNAs are shown as filled black symbols. In (D), the position of two pre-miRNA hairpins that did not pass the stringent filtering are indicated by asterisks marked with arrows. The analysis of the KSHV genome represents data reported in Grundhoff *et al.* (2006), from which (D) was partially reproduced (with permission). Analysis of the SV40 genome was performed with the current version of VMir (v1.5).

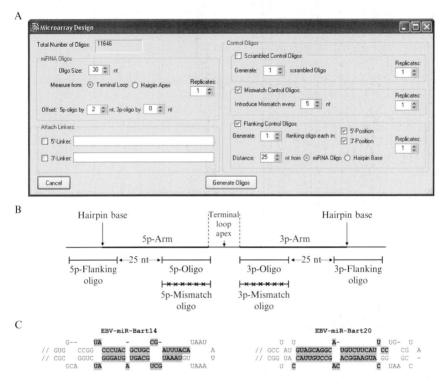

Figure 1.2 (A) Screen capture of VMir's microarray design tool. (B) Shows a linear representation of a hairpin to illustrate the general outline of oligo design according to VMir's default settings. The individual hairpin regions (5p-arm, terminal loop, and 3p-arm) are labeled, and the position of specific oligonucleotides (5p- and 3p-oligos) as well as default controls (5p-/3p-mismatch and flanking oligos) is indicated. (C) Predicted structures of two known, EBV-encoded pre-miRNA hairpins, which produce mature miRNAs from their 5p as well as 3p arms (only the terminal part of the structures is shown). The location of the mature miRNA molecules (shaded gray) within the precursors suggests that the terminal loop of miR-BART-14 was predicted correctly, whereas that of miR-BART-20 was predicted too small (instead, it is likely that all of the terminal 12 nt of miR-BART-20 form the terminal loop in the accurate structure).

hybridization results are finally matched back to the original hairpins. The protocol is meant as a guideline, and there are alternative routes the reader might take at any step instead of the described procedure. For example, rather than using VMir to design the arrays, one might simply use tiled arrays (it is more difficult to devise controls for such arrays, however, and it will be harder to match the results to individual hairpins). Our custom arrays were obtained from Combimatrix (www.combimatrix.com), but there are several other suppliers who provide custom arrays at reasonable cost (e.g., Agilent, www.agilent.com). We have used direct Cy3/Cy5 labeling of polyA-tailed RNA to generate probes for the array hybridization, but any of the numerous labeling methods described for miRNA microarray profiling can be used instead (see, e.g., Barad *et al.*, 2004). A caveat is that some labeling methods might not be compatible with the format of your particular custom array, and thus you should check with the supplier if in doubt. Naturally, the precise setup of the hybridization experiment will depend on the available material: We have primarily used two-color hybridizations with RNA from infected versus uninfected cells, but if you have no suitable uninfected control, you might use only one color or (as described later) hybridize miRNA-enriched versus non–miRNA-enriched RNA fractions from the same cells.

3.1. Array design

Equipment:

- Computer running VMir software (http://books.elsevier.com/companions/978012379179, requires Windows 2000, NT, or XP)

Install the software and perform an analysis following the instructions provided with the program. Open the result file using VMir Viewer, then select "Microarrays > Microarray Design..." from the "Tools" menu. This will open the dialog depicted in Fig. 1.2A. From this dialog you will have several choices for the design of oligonucleotides corresponding to the predicted hairpins as well as several types of control oligos. Figure 1.2B depicts the general outline of the oligo design according to the default design.

The software will design two specific oligos for each hairpin: one for the left and one for the right arm, covering the positions of putative 5p- and 3p-miRNAs. In animal pre-miRNA hairpins, mature miRNAs are typically located such that the 5p-miRNA ends at the terminal loop whereas the 3p-miRNA starts 2 nt into the hairpin stem (see the left part of Fig. 1.2C for an example). The software will therefore pick oligos that end or start (5p- or 3p-oligo, respectively) at the junction of hairpin stem and terminal loop. However, minimal free energy folding often predicts pre-miRNA hairpins with smaller terminal loop structures than those that are likely adopted *in vivo* (Zeng *et al.*, 2005). To account for such inaccuracies in the prediction (see the right part of Fig. 1.2C for an example) and ensure the complete miRNA is covered, the default size of the oligos is set to 30 nt (this setting is user-adjustable) instead of the typical size of a miRNA ($\sim$22 nt). It is also possible to offset the position of each oligo (positive values will shift the oligo toward the 3'-end; thus, 5p-oligos will move toward the terminal loop whereas 3p-oligos will move away from it). By default, the 5p-oligo is offset by 2 nt, whereas no offset is used for the 3p oligo; this corresponds in both cases to oligos that are extended by 2 nt toward the terminal loop, given that 3p-miRNAs typically start 2 nt into the stem. Alternatively, the user can chose to select oligos that simply end (5p-oligo) or start (3p-oligo) at the hairpin's apex position, but in this case significantly larger oligos should be chosen to account for the variable size of terminal loops.

There are three types of controls available from the design dialog (although our original array design did not contain any specific negative controls, these oligos will help to significantly reduce background noise). First, scrambled controls can be chosen for each of the specific oligos. While such oligos will account for unspecific hybridization resulting from high GC content, they will not control for cross-hybridization due to sequence similarity. Thus, by default the software will design mismatched instead of scrambled oligos. The user can select how frequently mismatches are introduced into the oligo. By default, a mismatch will be generated every 5 nt (as shown by Barad and colleagues [Barad *et al.*, 2004], as few as 2 mismatches are sufficient to eliminate specific miRNA binding). In addition to mismatch controls, the software will also generate flanking oligos. This control is intended to account for unspecific signals owing to, for example, small degradation products of viral transcripts which encompass the hairpin site. Such products will not be enriched in the region encoding the putative miRNA and therefore also should hybridize to the flanking control. By default, VMir picks one flanking oligo each in the 5'- and 3'-directions, at a distance of 25 nt from the 5p and 3p oligos. The distance is user-adjustable, and the user can also choose to calculate the distance from the base of the hairpin instead of the specific oligos. This is useful if there are concerns that the size selection procedure used to produce the hybridization material might not efficiently exclude pre-miRNAs (which are typically 60 to 80 nt in size), as these precursors might overlap with and hybridize to the flanking oligos. This will not be necessary if the gel-fractionation protocol described later or other stringent size selection methods are used. The user can also choose to generate more than one flanking oligo; if so, the chosen number of oligos will be arranged in a tiled fashion (with no overlap between adjacent oligos), starting from the selected position. There are cases in which a flanking control oligo might overlap with specific oligos from a neighboring hairpin; the control oligo will then be shifted farther away from the hairpin until the overlap is eliminated.

Finally, the user can attach custom linkers to the 5'- or 3'-end of each oligo. In our original design, we did not include any linkers, but others have shown that signal intensity decreases if the miRNA is located close to the 3'-end of an oligo (Barad *et al.*, 2004), possibly due to sterical hindrance because of attachment to the slide surface. It therefore might be preferable to attach a linker to the 3'-end of the oligos (we have not tested this ourselves). If using linkers, one should be aware of the maximum allowable length for the array oligos, as these values vary for the different array suppliers (e.g., 50 nt for Combimatrix, 60 nt for Agilent arrays).

Each oligo can be spotted once or repeatedly; the number of replicates in each category can be chosen by the user. A counter in the upper left corner of the dialog informs the user about the current total number of oligos so that the available array capacity (e.g., currently 12K for Combimatrix and up to 244K for Agilent arrays) can be optimally used. This counter also helps to ensure that the maximum number of spots is not exceeded. When using the default settings, a total of six oligos will be generated per hairpin (two specific ones, two mismatched, and two flanking controls). For large viral genomes, this might produce too many spots under conditions where no filter is used. For example, an analysis of the EBV genome yields a total of 3493 hairpins. This would make for a total of 20,958 oligos, a number that exceeds the maximum number of spots on some custom arrays (e.g., 12K Combimatrix arrays). However, most of these hairpins represent random "noise," and by setting a window count cutoff of four (an extremely mild filtering condition), only 1941 hairpins (or 11,646 oligos) remain, a number that will easily fit on a 12K custom array.

Once you hit the "Generate Oligos" button, the program will generate a tab-delimited text file with the names and sequences of the individual oligos. This text file can be directly submitted to the array supplier to order the custom arrays. The oligos are named by a string consisting first of an identifier for the hairpin (e.g., "MD1540") followed by a string identifying the type of oligo (e.g., "_5p," "_5pSc," "_5pMm," or "_5pFl" for specific, scrambled, mismatched, or flanking 5p-oligos, respectively) followed by an indicator for replicates, if any (e.g., "_R1" for the first replicate). Thus, the type and parental hairpin of a given hairpin can be simply determined by its name.

3.2. Microarray hybridization

This section describes the procedure that we have used to hybridize custom hairpin arrays. The protocol starts with total RNA (we employed total RNA isolated with RNA-Bee [Tel-Test Labs, Friendswood, TX] according to the manufacturer's instructions),which is fractionated on a denaturing polyacrylamide gel, eluted, aliquotted, and subsequently concentrated. To label the size-fractionated material for microarray analysis, the commercial MirVana miRNA labeling kit (Ambion, Austin, TX) is used. Section 3.2.2 provides a detailed, modified protocol that should be used as an additional reference in conjunction with the manufacturer's protocol. In this procedure, small RNAs are first "tailed" with Amine-coupled nucleotides with poly A polymerase and subsequently labeled with amino reactive Cy3/Cy5 dyes (Cyscribe, Amersham Biosciences). The labeled RNA is then hybridized to the custom arrays. As indicated previously, we have used Combimatrix 12K arrays together with the manufacturer's hybridization chamber and 1× hybridization miRNA microarray hybridization buffer (Ambion). Following hybridization, the arrays are washed and then scanned. We have used a GenePix 4000B Scanner (Molecular Devices Corp.) together with the GenePix Pro 6 software package. Subsequently, the data are further analyzed to identify miRNA candidates (described in Section 3.3).

3.2.1. Gel purification of small RNAs

Reagents (all should be nuclease-free):

- Nuclease-free water (Ambion, cat. # AM9937)
- 30% Acrylamide/Bis (29:1) (BioRad, cat. # 161-0156)
- TEMED (BioRad, cat. # 161-0800)
- 10% Ammonium Persulfate (APS)
- Linear acrylamide (5 mg/ml) (Ambion, cat. # AM9520)
- Gel loading buffer II (Ambion, cat. # AM8546G)
- MEGAclear kit (Ambion, cat. # AM1908)
- RNAse-free $1M$ NaCl
- TBE
- Urea
- Ethanol
- Eppendorf mini centrifuge tubes
- 15-ml conical tubes
- 1-ml syringe
- 5-ml syringe
- Rotator
- Vacuum
- 0.25″ Tubing
- Electric drill with 0.25″ bit

Pour a 15%, denaturing UREA, 0.5× TBE gel:

Set up mini gel casting tray. The gel apparatus used was a BioRad mini gel (8 × 10 cm) with 1.5-mm spacers and a 5-well comb. Mix the following reagents: 12.5 ml acrylamide, 2.5-ml 5× TBE, and 12 g urea. Bring with nuclease-free water up to 25 ml total, then heat to 55° (takes approximately 10 min) while periodically vortexing to completely dissolve the urea. Chill in ice water for approximately 10 min (to prevent premature polymerization while pouring the gel). Next, add 250 μl of 10% APS and 10 μl TEMED. Vortex briefly and pipette into casting apparatus. Allow to fully polymerize (takes ~30 min).

Run the gel:

Using 0.5× TBE, flush wells of gel with sterile needle and syringe. Fill the gel apparatus with 0.5× TBE. Load RNA using 0.5× to 1× gel loading buffer. Typically, 100 μg of RNA would be divided into three aliquots, which, after lyophilization, can be stored for later use at −80° for at least 3 months. Run gel at 250 V for approximately 20 min until bromphenol blue dye is 0.5 to 0.67 of the way down the gel.

Elute RNA from gel:

Using a sterile scalpel carefully cut out the band containing miRNA material. For routine array analysis, cut out a band from the top of the dark blue bromphenol blue dye to the middle of the xylene cyanol dye (ensuring that the cut out portion of the gel is identical for each sample to be analyzed). Using this strategy, we have repeatedly observed that the middle of the Xylene cyanol band migrates at ~35 nt, and the top bromphenol blue band migrates at ~10 nt. If smaller gel fragments are desired, an oligo nucleotide ladder combined with ethidium bromide staining can be used as a marker. (Note: We have found that hybridizing the miRNA-enriched gel purified size fraction [~15 to 25 nt] versus a non–miRNA-enriched size fraction [~25 to 35 nt] from the same sample/lane is an excellent method for distinguishing bona fide miRNA from minor degradation products of abundant transcripts. This is especially useful if an uninfected control sample is not available.) Place the gel slice into an RNAse-free mini centrifuge tube and carefully crush the gel slice using the plunger portion of a sterile 1-ml syringe. (Note: Great care must be taken at this step so as not to overflow the capacity of the tube.) After the gel is crushed into small (~3 mm^2) pieces, place the uncapped mini centrifuge tube upside down over an open 15-ml conical tube and, while holding the tubes together with one hand, force the gel slices into the conical tube by "flicking" the bottom of the upside-down mini centrifuge tube with you, other hand. (Note: If some pieces stick to the mini centrifuge tube, simply add a portion of the 1-M NaCl solution and repeat the process until all the gel fragments are removed.) Then add 10 ml of 1-M NaCl to tube with gel slice, cap, and rotate overnight at 4°. The next day, pellet the slurry at 2000×g. Decant into a 50-ml conical tube and store on ice, then add back 2 ml of fresh 10 ml NaCl to pelleted gel slurry. Vortex slurry for 2 h at room temperature, pellet as before, and decant 2 ml into 50-ml conical tube combining it with the initial 10 ml of eluate for 12 ml total.

miRNA Clean up:

(Note: Before you start, heat the MEGAclear elution solution to 95° for later use.) For the first time this procedure is to be performed drill a hole in the bottom of a 15-ml conical tube large enough to pass the tubing from the aspirator through the bottom to the top of the 15-ml conical tube (Note: Save the tube for reuse in future miRNA preps.) Then, attach the tubing to the bore of a 5-ml syringe, which rests in the conical tube (this serves as holder for the MEGAclear column which fits snugly into the top of the syringe). Add 1.5 ml of linear acrylamide (5 mg/ml, Ambion) and 18 ml ethanol (60% final V/V) and vortex to mix thoroughly, then apply mixture to MEGAclear filter cartridge by utilizing the vacuum to pull the mixture through the filter. This takes multiple rounds of pipetting ~1 ml at a time into the MEGAclear cartridge. Afterward, "clean" the precipitated RNA by adding 4 ml of 80% ethanol and allowing the vacuum to pull it through the column. Next, place the MEGAclear cartridge into a collection tube and add 500 ml of 80% ethanol. Spin for 1 min at 5000×g in a mini centrifuge. Then, dry completely spinning at 10,000×g for 1 min to remove any residual ethanol. Remove the MEGAclear column and place into a fresh collection tube.

miRNA elution and concentration:

Add 50 μl of heated (95°) elution solution directly to matrix and let site for 2 min. Centrifuge at 10,000×g for 1 min and repeat the process for a final eluate of 100 μl. Check optical density (OD) of undiluted sample to ensure presence of RNA, aliquot into three separate tubes, and then dry completely in a vacuum concentrator set to medium heat. Samples are now ready for the MirVana labeling reaction or storage at −80°.

3.2.2. Labeling and array hybridization

Reagents:

- MirVana miRNA labeling kit (Ambion, cat. # AM1562)
- Cyscribe Cy3 and Cy5 dyes (Amersham, cat. # 25-8010-80 and 25-8010-79; or cat. # RPN 5661)
- miRNA hybridization buffer (Ambion, cat. # AM8860G)
- Detergent concentrate (Ambion, cat. # AM9823G5)
- Salt concentrate (Ambion, cat. # AM9763G6)
- Incubator pre-equilibrated to 65°
- Mini centrifuge tubes
- Custom microarray (e.g., Combimatrix 12K)
- Microarray scanner (e.g., Axon 4000B Scanner or similar)
- Microarray analysis software (e.g., Genepix 6 pro or similar)

Probe labeling:

Resuspend aliquot of purified small RNAs in 4 μl of nuclease-free water and mix by vortexing. Set up tailing and dye coupling reactions exactly as suggested by manufacturer's directions and incubate for 1 h in a dark drawer to protect from light-induced quenching. While the reaction is incubating, preheat the elution buffer at 95° in preparation for the elution step. Clean up the unincorporated dye using the kit spin columns and wash buffer. Elute by adding 40 μl of elution buffer directly to the matrix and place immediately in the 65° incubator for 5 min. (Note: This is a deviation from the manufacturer's suggested protocol using an increased volume of elution buffer because the hybridization chambers for Combimatrix arrays have a larger capacity than arrays that hybridize with lifter cover slips.) Spin reaction at 10,000×g for 1 min, then repeat the elution with an additional 40 μl using 95° elution buffer and 5 min incubation at 65°, and pool with original eluate for a total of ∼75 μl. (Note: We typically lose ∼10% of input volume, so we have designed the protocol to have extra capacity to ensure at least 67 μl which is required for the next hybridization step.)

Array hybridization:

Assemble the array according to the manufacturer's (Combimatrix) suggestions, or, alternatively, two medium binder clips can be used to hold the hybridization chamber in place on the array. Prehybridize the array with $1\times$ hybridization for 2 min; while this is occurring, mix the $3\times$ hybridization buffer with 67 μl of the labeled small RNAs to a final of $1\times$ and heat to 95° for 3 min. Remove the prehybridization solution from the array and immediately add the prewarmed, labeled RNA hybridization solution. Wrap in foil and place in rotisserie hybridization oven by securing in place with tape, and then rotate at 42° overnight.

Washing array:

The next day, make 1 ml each of low (940-μl nuclease-free water, 10-μl detergent concentrate, 50-ml salt concentrate) and high stringency (995-μl nuclease-free water, 5-μl salt concentrate) wash buffers. Remove the array from the oven and foil and quickly rinse with low stringency wash buffer by pipetting up and down in the chamber. Repeat this process at least 10 times with 100 μl of fresh wash buffer each time for ~2 min total wash time. Repeat the process with the high-stringency wash buffer. Add 100 μl of imaging solution and dismantle the hybridization chamber leaving the imaging solution on the array surface. Cover with a lifter cover slip by carefully starting from one corner and allowing the rest of the cover slip to sit down, taking great care to not get any air bubbles. Place Kim wipes on top of cover slip and gently apply pressure to remove any excess imaging solution that escaped the cover slip. Scan arrays according to the manufacturer's instructions (e.g.,Combimatrix arrays should be scanned wet) and proceed to data analysis.

3.3. Data analysis

Once you have scanned your arrays, the obtained data must be analyzed to identify the hairpins with the highest probability of representing pre-miRNAs. Owing to the design of the array (i.e., individual control spots for every specific oligo), it is not possible to completely perform this analysis using commercial array analysis software (e.g., Genepix 6 will allow you to identify control oligos based on pattern-matching the oligo name, but all oligos matching the pattern together will be used as a pool of control spots for all other spots on the array). Therefore, VMir incorporates a tool that reads the results generated by the primary array scanning/analysis software, then matches the oligos to the original hairpins and integrates the results (Fig. 1.3). VMir is natively able to read result files generated by Genepix (gpr files), but it accepts any tab-delimited text file. Nearly every array analysis software will allow you to export the scanned data in such a format (you will have to consult your manual for specific instructions). Detailed instructions regarding the required format for these files is provided with the program.

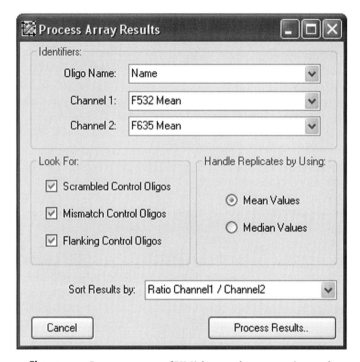

Figure 1.3   Screen capture of VMir's array data processing tool.

To open the tool, select "Microarrays > Process Microarray Data..." from the "Tools" menu. The data columns that VMir should use to look for the oligo name and fluorescence intensities (e.g., Cy5 and Cy3) are identified by entering the appropriate header text in the three text fields labeled "Oligo Name," "Channel 1," and "Channel 2." By default, these are set to Genepix-specific values: the header text for the oligo name is "Name," channel 1 is set to "F532 Mean" (mean signal of the Cy3 channel without background correction), and channel 2 is set to "F635 Mean" (mean signal of the Cy5 channel without background correction). Note that, when analyzing Combimatrix arrays with Genepix, you should not use background-corrected values, as the semiconductor array grid will throw the correction off. However, when using other arrays and background correction is desired, you can select the corresponding data column header from a dropdown menu (e.g., selecting "F532 Mean — B532" will use the background corrected mean Cy3 signal). If you are not using Genepix as the primary analysis software, you will have to enter the appropriate column headers yourself (see instructions provided with VMir).

After reading the results, the program will first identify the oligos based on the naming scheme described in Section 3.1 (it is thus important that these names are not modified) and match them to the individual hairpins. If more than one replicate of a given oligo is found, either representative mean or median values are calculated, depending on the user's choice (by default, this is set to mean values). Among the control types, the types with the highest hybridization signal is then chosen, and its value is subtracted from that of the specific oligo to obtain a corrected value. By default, the program searches for all types of control oligos (scrambled, mismatched, and flanking), but the user can choose which of the categories should be employed. By default, the program subsequently calculates the ratio of the first versus the second channel as the final score, but the user can also choose to calculate the inverse ratio or use only a single channel. Hairpins are then awarded the score of either their 5p- or 3p-oligo, whichever is greater. The results of the analysis are subsequently written to a file which lists the hairpins, sorted according to their score, along with the original and corrected values of the specific 5p- and 3p-oligos. The file is in a standard, tab-delimited text format, which can be opened, annotated, edited, and re-sorted in other applications, such as Excel. Based on these results, the top scoring candidates can then be chosen for final confirmation by Northern blot detection of small RNAs.

4. Concluding Remarks

Computational prediction of miRNAs is increasingly used to identify novel miRNA candidates. Many prediction algorithms have been reported, but because executable files are rarely provided, these methods are of limited use for researchers who do not have sufficient programming knowledge to implement these algorithms themselves. We provide a protocol that uses VMir, a freely available program, to identify potential pre-miRNA candidates within relatively small sequences (such as viral genomes) and test these candidates using microarrays. The method is especially advantageous for the analysis of large viral genomes because many predictions can be tested at once, and we consider it a valuable complement to existing cloning methods that might be biased against certain miRNAs (Berezikov *et al.*, 2006; Elbashir *et al.*, 2001). We stress, however, that it should not be used exclusively: First, the method has relatively high technical demands, and cases where no positive result is obtained might be difficult to interpret. Second, it is possible that array analysis has its own biases (i.e., certain miRNAs might be notoriously difficult to pick up on an array), and thus a combination of both methods appears optimal. Lastly (and perhaps most importantly), cloning determines the precise 5'- and 3'-ends of a given miRNA, information that our protocol is unable to provide. Thus, if in doubt, we recommend conducting a cloning experiment first before proceeding to the type of analysis described here. In these cases, however, the given protocol will enable researchers, including those with limited computer literacy, to conduct a thorough analysis of viral genomes based on computational prediction and microarray analysis.

REFERENCES

Ambros, V., Bartel, B., Bartel, D. P., Burge, C. B., Carrington, J. C., Chen, X., Dreyfuss, G., Eddy, S. R., Griffiths-Jones, S., Marshall, M., Matzke, M., Ruvkun, G., and Tuschl, T. (2003). A uniform system for microRNA annotation. *RNA* **9**, 277–279.

Barad, O., Meiri, E., Avniel, A., Aharonov, R., Barzilai, A., Bentwich, I., Einav, U., Gilad, S., Hurban, P., Karov, Y., Lobenhofer, E. K., Sharon, E., Shiboleth, Y. M., Shtutman, M., Bentwich, Z., and Einat, P. (2004). MicroRNA expression detected by oligonucleotide microarrays: System establishment and expression profiling in human tissues. *Genome Res.* **14**, 2486–2494.

Berezikov, E., Cuppen, E., and Plasterk, R. H. (2006). Approaches to microRNA discovery. *Nat. Genet.* **38**(Suppl.), S2–S7.

Burnside, J., Bernberg, E., Anderson, A., Lu, C., Meyers, B. C., Green, P. J., Jain, N., Isaacs, G., and Morgan, R. W. (2006). Marek's disease virus encodes MicroRNAs that map to meq and the latency-associated transcript. *J. Virol.* **80**, 8778–8786.

Cai, X., Lu, S., Zhang, Z., Gonzalez, C. M., Damania, B., and Cullen, B. R. (2005). Kaposi's sarcoma-associated herpesvirus expresses an array of viral microRNAs in latently infected cells. *Proc. Natl. Acad. Sci. USA* **102**, 5570–5575.

Cai, X., Schafer, A., Lu, S., Bilello, J. P., Desrosiers, R. C., Edwards, R., Raab-Traub, N., and Cullen, B. R. (2006). Epstein-Barr virus microRNAs are evolutionarily conserved and differentially expressed. *PLoS Pathog.* **2**, e23.

Cantalupo, P., Doering, A., Sullivan, C. S., Pal, A., Peden, K. W., Lewis, A. M., and Pipas, J. M. (2005). Complete nucleotide sequence of polyomavirus SA12. *J. Virol.* **79**, 13094–13104.

Cui, C., Griffiths, A., Li, G., Silva, L. M., Kramer, M. F., Gaasterland, T., Wang, X. J., and Coen, D. M. (2006). Prediction and identification of herpes simplex virus 1-encoded microRNAs. *J. Virol.* **80**, 5499–5508.

Cullen, B. R. (2006). Viruses and microRNAs. *Nat. Genet.* **38**(Suppl.), S25–S30.

Dunn, W., Trang, P., Zhong, Q., Yang, E., van Belle, C., and Liu, F. (2005). Human cytomegalovirus expresses novel microRNAs during productive viral infection. *Cell Microbiol.* **7**, 1684–1695.

Elbashir, S. M., Lendeckel, W., and Tuschl, T. (2001). RNA interference is mediated by 21- and 22-nucleotide RNAs. *Genes Dev.* **15**, 188–200.

Grey, F., Antoniewicz, A., Allen, E., Saugstad, J., McShea, A., Carrington, J. C., and Nelson, J. (2005). Identification and characterization of human cytomegalovirus-encoded microRNAs. *J. Virol.* **79**, 12095–12099.

Grundhoff, A., Sullivan, C. S., and Ganem, D. (2006). A combined computational and microarray-based approach identifies novel microRNAs encoded by human gamma-herpesviruses. *RNA* **12**, 733–750.

Gupta, A., Gartner, J. J., Sethupathy, P., Hatzigeorgiou, A. G., and Fraser, N. W. (2006). Anti-apoptotic function of a microRNA encoded by the HSV-1 latency-associated transcript. *Nature* **442**, 82–85.

Hofacker, I. L., Fontana, W., Stadler, P. F., Bonhoeffer, S., Tacker, P., and Schuster, P. (1994). Fast folding and comparison of RNA secondary structures. *Monatshefte f. Chemie* **125**, 167–188.

Omoto, S., and Fujii, Y. R. (2005). Regulation of human immunodeficiency virus 1 transcription by nef microRNA. *J. Gen. Virol.* **86,** 751–755.

Omoto, S., Ito, M., Tsutsumi, Y., Ichikawa, Y., Okuyama, H., Brisibe, E. A., Saksena, N. K., and Fujii, Y. R. (2004). HIV-1 nef suppression by virally encoded microRNA. *Retrovirology* **1,** 44.

Pfeffer, S., and Voinnet, O. (2006). Viruses, microRNAs and cancer. *Oncogene* **25,** 6211–6219.

Pfeffer, S., Lagos-Quintana, M., and Tuschl, T. (2003). Cloning of small RNA molecules. *In* "Current Protocols in Molecular Biology" (F. M. Ausubel, R. Breut, R. E. Kingston, D. D. Moore, J. G. Seidman, J. A. Smith, and K. Struhl, eds.), Vol. 4, pp. 26.4.1–26.4.18. John Wiley & Sons, New York.

Pfeffer, S., Zavolan, M., Grasser, F. A., Chien, M., Russo, J. J., Ju, J., John, B., Enright, A. J., Marks, D., Sander, C., and Tuschl, T. (2004). Identification of virus-encoded micro-RNAs. *Science* **304,** 734–736.

Pfeffer, S., Sewer, A., Lagos-Quintana, M., Sheridan, R., Sander, C., Grasser, F. A., van Dyk, L. F., Ho, C. K., Shuman, S., Chien, M., Russo, J. J., Ju, J., *et al.* (2000). Identification of microRNAs of the herpesvirus family. *Nat. Methods* **2,** 269–276.

Samols, M. A., and Renne, R. (2006). Virus-encoded microRNAs: A new chapter in virus-host cell interactions. *Fut. Virol.* **1,** 233–242.

Samols, M. A., Hu, J., Skalsky, R. L., and Renne, R. (2005). Cloning and identification of a microRNA cluster within the latency-associated region of Kaposi's sarcoma-associated herpesvirus. *J. Virol.* **79,** 9301–9305.

Sullivan, C. S., and Ganem, D. (2005). MicroRNAs and viral infection. *Mol. Cell* **20,** 3–7.

Sullivan, C. S., Grundhoff, A. T., Tevethia, S., Pipas, J. M., and Ganem, D. (2005). SV40-encoded microRNAs regulate viral gene expression and reduce susceptibility to cytotoxic T cells. *Nature* **435,** 682–686.

Zeng, Y., Yi, R., and Cullen, B. R. (2005). Recognition and cleavage of primary microRNA precursors by the nuclear processing enzyme Drosha. *EMBO J.* **24,** 138–148.

CHAPTER TWO

ROBUST MACHINE LEARNING ALGORITHMS PREDICT MICRORNA GENES AND TARGETS

Pål Sætrom[*,†,§] *and* Ola Snøve, Jr.[*,‡,§]

Contents

[*]Interagon AS, Laboratoriesenteret, Trondheim, Norway
[†]Department of Computer and Information Science, Norwegian University of Science and Technology, Trondheim, Norway
[‡]Department of Cancer Research and Molecular Medicine, Norwegian University of Science and Technology, Trondheim, Norway
[§]Division of Molecular Biology, Beckman Research Institute of the City of Hope, Duarte, California

Methods in Enzymology, Volume 427
ISSN 0076-6879, DOI: 10.1016/S0076-6879(07)27002-8

Abstract

MicroRNAs (miRNA) are nonprotein coding RNAs with the potential to regulate the gene expression of thousands of protein coding genes. Current estimates suggest the number of miRNA genes may be twice of what is currently known, and the mechanisms governing miRNA targeting remain elusive. Machine learning algorithms can be used to create classifiers that capture the characteristics of verified examples to determine whether genomic hairpins are similar to verified miRNA genes or if message 3′UTRs possess known target characteristics. Algorithms can never replace biological verifications, but should always be used to guide experimental design. This chapter focuses on potential problems that must be addressed when machine learning is used and follows a practical approach to demonstrate how support vector machines and genetic programming can predict miRNA genes and targets.

1. INTRODUCTION

When miRNAs were discovered in 2001, the most convincing argument for their general importance was that these nonprotein coding genes appeared to be highly conserved throughout evolution (Lagos–Quintana et al., 2001; Lau et al., 2001; Lee and Ambros, 2001). A class of sequences that are conserved from worms and flies to rodents and humans is likely to be functional in some way or another. Indeed, miRNAs have the potential to cleave (Yekta et al., 2004), degrade (Bagga et al., 2005; Wu et al., 2006), or suppress translation (Olsen and Ambros, 1999; Reinhart et al. 2000) of thousands of genes (Lewis et al., 2005; Lim et al., 2005). This potential for regulation of more than one third of, and perhaps, all human genes has spurred a tremendous interest in miRNAs. This interest is also fueled by the putative involvement of miRNAs in cancer (Esquela-Kerscher and Slack, 2006) and great diagnostic (Lu et al., 2005) and therapeutic (Behlke, 2006) potential.

So what made the initial detection of numerous miRNAs possible? After all, miRNAs had gone unnoticed for decades despite abundant expression in many extensively studied cell lines. The answer is painfully obvious: Scientists knew what they were looking for due to chance discoveries of lin-4 and let-7—two nonprotein coding RNAs that had already been demonstrated to be important for larval development in *Caenorhabditis (C.) elegans* (Olsen and Ambros, 1999; Reinhart et al., 2000; Wightman et al., 1993). When lin-4 was found in the early nineties, it had neither paralogs in the worm nor orthologs in other species, but let-7, which was discovered 7 years later, is well-conserved in multiple species (Pasquinelli et al., 2000). Because evolutionary pressure clearly existed to preserve let-7, this supported the idea that the silencing mechanism was more general than previously thought. The characteristics of the two founding members of the miRNA class were used to design experiments that made everyone realize lin-4 and let-7 were the tip of an iceberg. In a way, these scientists constructed a simple algorithm to look for new RNAs that were similar to the two already known.

One should always take advantage of existing data to create algorithms that can guide research and maximize the chances of success. This may be an obvious statement, but accomplishing it may often be difficult for at least two reasons. First, biology research moves fast and data, information, and knowledge are frequently changed and updated. Consequently, algorithms have to be updated relatively often. Second, many biological problems cannot be adequately modeled by simple linear algorithms, but require complex nonlinear models. Machine learning algorithms can automatically build such models from example data and have therefore become popular for solving many problems in computational biology. Examples include finding remotely homologous proteins (Jaakkola *et al.*, 2000), predicting promoters (Bajic *et al.*, 2004) or exon–intron splice sites (Hebsgaard *et al.*, 1996), finding microarray expression patterns common to specific cancers (Khan *et al.*, 2001), and more. Here, we will show how to use machine learning to separate miRNA genes from random genomic hairpins and miRNA target sites from nontargeted $3'$ UTR regions.

A practical approach is taken throughout the chapter. First, we discuss some machine learning caveats, outline some methodological considerations, and present a guide to the various steps of a typical machine learning experiment. Second, we use one popular algorithm, the support vector machine (SVM) of statistical machine learning, to create a classifier for miRNA gene prediction. Finally, for the sake of illustration, we show how another algorithm with similar statistical properties, namely boosted genetic programming (GP), can predict miRNA target sites.

2. APPROPRIATE USE OF MACHINE LEARNING

Machine learning approaches can roughly be divided into unsupervised and supervised algorithms. Unsupervised algorithms comprise the various clustering schemes and are appropriate when you want to group data about which you have limited or no prior information (D'Haeseleer, 2005). An unsupervised algorithm could, for instance, be useful if you wanted to group known miRNAs into a number of unknown families. In this chapter, however, we will only be concerned about supervised learning algorithms—that is, we will assume that the groups have already been assigned and that the task is to find characteristic features that separate them. Supervised machine learning encompasses algorithms for both regression and classification, but we will only treat two-way or binary classification problems (i.e., separating class A from class B). miRNAs fit into this class of problems, as a genomic hairpin is either a miRNA, or it is not—similarly, a $3'$ UTR site is either a target for a given miRNA, or it is not.

2.1. Data sets

The goal for every supervised learning method, be it artificial neural networks (Dayhoff and DeLeo, 2001), decision trees (Safavian and Landgrebe, 1991), naïve Bayes classifiers (Domingos and Pazzani, 1997), SVMs (Burges, 1998), or boosted learners (Meir and Rätsch, 2003), is to achieve the highest possible performance on unseen data. It always requires the assembly of a training set, which ideally would comprise information about all types of examples within each group. More specifically, ideal training sets should contain samples that are mutually independent and are true representatives of what we are trying to classify. Using statistical terms, the samples should be independent and identically distributed. This also means that the training set should not contain mislabeled instances, but, in practice, algorithms must be able to handle some noise, as it is impossible to construct ideal training sets. For instance, as neither all miRNAs nor target sites are known, we do not know whether current databases accurately represent all miRNA families or target sites with various characteristics. Erroneous labeling is also entirely possible, as annotated miRNAs may belong to other, possibly unknown families of expressed RNAs. Furthermore, annotated targets may not be endogenously relevant, as down regulation could be contingent on gross overexpression of miRNAs or be the secondary result of altered expression among the actual targets.

It is important to keep in mind that the algorithm should be optimized for performance on unseen data. A superior predictor with high performance on unseen data always does well on the training set, but the converse is not always true. Many learning algorithms can achieve arbitrarily high performance on their training set by increasing the complexity of the classifier, but the performance may deteriorate on unseen data. The phenomenon is known as overfitting, and is shown schematically in Fig. 2.1. An important consideration when using machine learning algorithms is therefore to estimate a classifier's true performance on unseen data. This is often referred to as the classifiers generalization error or performance. The following sections will discuss performance estimation and performance measures in more detail and outline a stepwise procedure for how machine learning algorithms can be used to solve classification problems.

2.2. Performance estimation

The simplest method to estimate a classifier's performance is to leave a sizable fraction of the samples out of the training set, and use the trained classifier's performance on this test set to estimate its performance. Theoretically, as

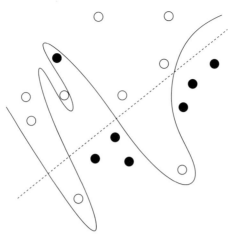

Figure 2.1 Schematic depiction of overfitting. The solid line is able to separate the white from the black balls, and is therefore a perfect classifier on this data set. The broken line's performance is, however, more likely to hold up when presented with new examples.

the algorithm did not see the examples during training, the performance on the test set would be a good estimate for the classifier's true performance. Constructing a representative test set is not always possible, however, as many problems amenable to machine learning have limited data available. Thus, creating a large separate test set may prevent the algorithm from learning all aspects of the available data. In such cases, various forms of bootstrapping are used to estimate the performance (Martin and Hirschberg, 1996; Stone, 1974). One simple approach is to create 10 equally sized, randomly generated subsets of the data, train one classifier for every combination of nine subsets, and estimate the performance by reporting the average performance these classifiers have on the 10th subset that they were not trained on. The approach is called 10-fold cross-validation. Another popular method is leave-one-out cross-validation, which involves the same training of several classifiers and reporting of averaged test set performance, but the classifiers are trained on all but one sample instead of all but one subset.

To get a true estimate of the classifier's performance, it is pivotal that the training and test sets are independent. More specifically, if the initial data set contains any instances that are closely related, such as several miRNAs from the same family, all of these instances should be placed in the training or test set. Otherwise, one risks getting biased performance estimates, as a classifier will more likely recognize instances that are closely related to the ones in its training set, such as miRNAs from the same family, than it will recognize independent instances, such as miRNAs from a different family.

A similar caution goes for parameter optimization. Most machine learning algorithms have several input parameters that users can fine-tune for optimal performance on a particular problem. Using the test set to optimize these parameters will, however, risk overfitting the classifier to the test set, thereby producing biased performance estimates (Salzberg, 1997). To get unbiased estimates, users must therefore keep parameter optimization independent of the final test set.

In this chapter, we will demonstrate two types of performance estimation by using 10-fold cross-validation for miRNA gene prediction and leave-one-out cross-validation in the miRNA target prediction experiments. This reflects that fewer validated examples are available for the target prediction than for gene prediction, and we are therefore unable to divide the samples into the representative subsets that 10-fold cross-validation requires. Furthermore, we will demonstrate how improper division of the dataset into training and test sets can lead to biased performance estimates.

2.3. Performance measures

The previous section described classifier performance, but it did not define how performance should be measured. Importantly, classifier performance can be measured in many different ways (Baldi *et al.*, 2000). We will introduce some of these measures but must first define some common quantities that pertain to predictions. No matter which algorithm is chosen, the final classifiers will get some predictions right and some wrong. In the miRNA gene prediction example, if a classifier predicts a miRNA and it is later verified, it is a true positive (TP), whereas it would be a false positive (FP) if it turned out not to be a real miRNA. Conversely, a genomic hairpin correctly predicted not to be a miRNA is a true negative (TN), whereas it would be a false negative (FN) if the same hairpin was later verified to be a Drosha and Dicer substrate.

Several performance measures can be defined from the aforementioned quantities. For example, the sensitivity (Se) is defined as the fraction of positive cases correctly predicted by the algorithm (Se = TP / [TP + FN]) — in our example, the sensitivity equals the fraction of miRNAs correctly predicted by the algorithm. Achieving perfect sensitivity is easy; however, just say that all genomic hairpins are miRNAs and the classifier hits the rooftop as far as sensitivity goes. Of course, it would be a lousy classifier despite the high sensitivity, as it would get all the negative examples wrong. Consequently, a classifier's specificity (Sp) must be reported to show how well it performs on the negative examples (Sp = TN / [TN + FP]). Returning to the miRNA gene prediction example, specificity is the fraction of non–miRNA genomic hairpins correctly identified by the classifier.

The algorithms mentioned in this chapter score every sample. A higher score means a higher probability of being a miRNA or a target, which means that a score threshold must be set to determine whether a sample's score is high enough to be labeled a candidate miRNA gene or target. A low score threshold will give a high sensitivity at the expense of the specificity and vice versa. It is therefore beneficial to observe the relationship between sensitivity and specificity over the full range of possible thresholds to obtain an appropriate classifier. A receiver operator characteristic (ROC) curve does just that. Because the highest sensitivity is achieved at the lowest specificity and vice versa, it is customary to plot the reminder of one fraction to get a plot that starts at the origin and approaches unity (TP rate versus FP rate as shown in Fig. 2.5B). A random classifier would have a ROC curve that was a straight line from the lower left to the upper right corner. To compare classifiers, the area under the ROC curve is often a good measure, but other measures are more relevant when predictions are used to guide biologists in experimental design. For example, if the only important parameter is to find one or a few correct predictions, it may be better to consider the positive predictive value (PPV = TP / [TP + FP]), which in the miRNA example would be the fraction of predicted miRNAs that turned out to be correct.

Note that the measures we have given so far depend on dataset characteristics. For example, if we create a lot of negative examples—when we get to miRNA gene prediction, we will start with millions of genomic hairpins—it will be biased toward higher sensitivity and lower specificity. It is therefore not necessarily possible to compare these performance measures across datasets. When we do so we must ensure that the datasets are comparable, and even then be aware that what is obtained is merely an indication. For that reason, it is useful to optimize a single measure that does not depend on the ratio of positive to negative examples in one particular dataset. We use the Pearson correlation coefficient, but various measures of correlation can be used for this purpose.

2.4. Machine learning steps

To summarize this section, the steps one would commonly follow when using machine learning to solve classification problems are outlined. The steps are:

(1) decide on input;
(2) assemble training set;
(3) train classifier and estimate its performance; and
(4) evaluate classifier predictions.

Later in the chapter, these steps show how to predict miRNA genes and target sites.

Deciding on input means determining which features or variables the machine learning algorithm will see and base its predictions on. Ideally, the input should be limited to the features relevant to the problem, but this is often impossible in practice. For some applications, such as microarray analyses, finding the relevant features is part of the problem. For other problems, such as miRNA gene or target prediction, the user may not know all relevant features. Consequently, one often has to settle for a relatively big set of putatively important features, such as all the probes on a microarray, several primary or secondary structure attributes of miRNA hairpins, or the miRNA target site sequence.

Regardless of which input features are selected, a training set of representative positive and negative input examples needs to be assembled. As previously mentioned, some mislabeled examples must be expected. Consequently, robust machine learning algorithms that handle noise in the training set are essential when training the classifiers. Finally, the user should evaluate the trained classifier and its predictions. This evaluation should always compare the classifier's performance to that of existing classifiers. Classifiers with improved performance over existing solutions are clearly useful, but classifiers with similar or even worse performance are valuable if their predictions complement existing predictions. To illustrate, consider the Venn-diagrams in Fig. 2.2. On one hand, if classifier A has more TP and fewer FP predictions than classifier B and the TP and FP predictions are subsets and supersets of the other classifier's predictions, then classifier A is better than classifier B (Fig. 2.2; left example). On the other hand, if there is little or no overlap between the predictions of classifiers A and B (Fig. 2.2; right example), both have their own merit and can potentially be combined to create a classifier with superior performance. In any case, one should always consider predictions supported by multiple classifiers and other independent data as primary candidates for further lab verifications.

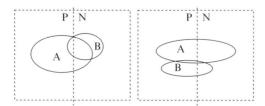

Figure 2.2 Example of noncomplementary and complementary classifiers. The ratio of the ellipses' areas to the total area on either side of the middle vertical line shows the fraction of true positive and true negative predictions. When classifiers A and B overlap, it means that they both get this part of the dataset right. In the left example, classifier B does not contribute true positive predictions in addition to those of classifier A, nor does B predict more true negatives than A, which means that A is the better classifier. In the right example, the classifiers perform well on different parts of the dataset, which means that they can both be valuable.

3. One Deterministic and One Stochastic Algorithm

Supervised machine learning algorithms, of which there are many, can be divided into two main categories. Deterministic algorithms will always end up with the same classifier given a set of input parameters and a training set, whereas classifiers obtained from stochastic algorithms may differ between independent runs. Here, we will use SVMs, a deterministic method, to predict miRNA genes, and boosted GP, a stochastic method, to identify miRNA targets. Although the algorithms will be outlined in the following, a detailed explanation requires extensive use of a mathematical framework and is considered out of the scope of this chapter. Interested readers should consult specific reviews for good introductions to SVMs (Burges, 1998), genetic programming (Banzhaf *et al.*, 1997), and boosting (Meir and Rätsch, 2003), but proper use is more important than detailed knowledge of the algorithms, especially for those more interested in the application than the technical details of the algorithm.

3.1. Support vector machines

Support vector machines were popularized in the late nineties. Their theory is appealing, at least for the mathematically inclined, because error bounds can be computed, thus providing a lower limit for the algorithm's performance (Vapnik, 1998). For now, however, the bounds are not practical—they are similar to saying more than one, less than a billion—but the algorithms have nevertheless proven themselves empirically.

Support vector machines operate on numerical input vectors, which means that biological data, be it sequences, structures, or other putatively important features, must be mapped to numerical values. Needless to say, this mapping plays an important role for the ultimate performance in the SVM. Hence, when SVMs are said to be deterministic, it should not be taken to imply that all SVMs produce the optimal classifier independent of the user's choices of several parameters, of which the vector mapping is only one.

Based on the input vectors and their assigned class membership, however, all SVMs try to find the maximum margin hyperplane—the mathematical generalization of points in one dimension, lines in two dimensions, and planes in three dimensions—that separates the two classes. The maximum margin hyperplane is the one with maximum distance to the closest sample, as illustrated with a two-dimensional (2D) example of a separable problem in Fig. 2.3A. The desire for linear classifiers—or higher dimensional generalizations of such—comes from a much higher potential for overfitting with more complex classifiers. Of course, a straight line cannot

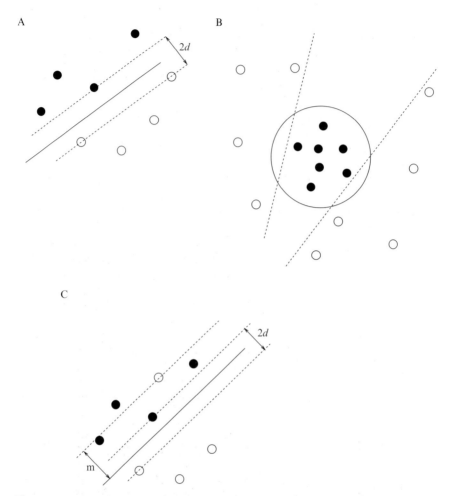

Figure 2.3 Two-dimensional examples SVM classifiers. (A) The problem is linearly separable when all the black balls can be separated from the white balls by a straight line. A maximal margin classifier is then the straight line that has the maximum distance to the closest example; here, the margin is d, and the distance between the closest black and white balls becomes $2d$. (B) Broken lines are clearly not able to do a good job in this example, whereas a more complex classifier—in this case a circular one—is a good choice. Support vector machines achieve this by a mathematical trick that allows the algorithm to operate with the equivalent of a two-dimensional straight line in higher dimensions. (C) All training sets are not separable, neither in the input space nor in the higher dimensional space, which is why SVMs can allow some misclassifications. While still maximizing the margin, d, a soft margin classification algorithm also minimizes the cumulative distance to all misclassified examples, which in this case is just one white ball at distance m from the classifier.

always separate one class of examples from the other. Some problems require nonlinear solutions, whereas mislabeled instances or other noise in the training set can turn a linearly separable problem into a problem that seemingly requires more complex solutions. Support vector machines use two means to solve these two problems. Kernel mapping lets the SVM construct nonlinear classifiers to solve more complex problems (Fig. 2.3B), whereas soft margins allow instances in the training set to be misclassified, thereby limiting the potential for overly complex classifiers (Fig. 2.3C).

The kernel mapping method is a mathematical trick, in which an SVM maps the input vectors to a higher dimensional space using a so-called kernel function and then attempts to find a hyperplane classifier in that space instead. The linear classifier in higher dimensions will correspond to a nonlinear classifier in the input space. Consequently, SVMs achieve non-linearity, but always operate with higher dimensional generalizations of a straight line. Even though the mapping violates the theory that ensures the generalization performance and error bounds of SVMs, a growing body of evidence suggests the resulting classifiers will indeed perform very well on unseen data compared with other algorithms. It should also be noted that other mathematical tricks allow SVMs to avoid the potentially time-consuming mapping to higher dimensions, but rely on kernel computations directly on the input vectors. Consequently, the run time for a certain experiment will only depend on the number of samples and the dimension of the input vectors.

Soft margins, like the standard maximum margins, maximize the distance to the closest correctly classified sample. In addition, soft margins minimize the cumulative distance to all erroneously classified samples (Fig. 2.3C). As a result, soft margin classifiers find the optimal tradeoff between correct and wrong classifications, and one can use this to control for noise in the training set.

The downside to kernel mappings and soft margins is that both methods require extra parameters, as users have to choose suitable kernel functions, some of which also require additional parameters, and the appropriate tradeoff between correct and wrong classifications for the soft margins. When the parameters, including the input features and the kernel function, have been chosen, an SVM will, however, provide the same output every time.

3.2. Boosted genetic programming

Whereas SVMs are deterministic, GP is stochastic, which means that it depends on random events (Koza *et al.*, 1999). Consequently, independent GP runs will not yield identical classifiers. Genetic programming may therefore require multiple runs to find a classifier that satisfies the desired performance characteristics.

A typical GP procedure starts by randomly generating a population of solutions. Individual solutions are then assigned a fitness score based on their ability to solve the problem, and the fit individuals—we measure fitness with the Pearson correlation—are allowed to sporadically change (mutate) and exchange (crossover) parts of their solution representation (genome). In the case of miRNA target sites, the solutions are sequence motifs that separate between validated targets and a set of sequences that are perceived not to be targets. As shown in Fig. 2.4, motifs can be represented as tree structures in which evolution can be introduced by changing tree elements and exchanging subtrees to simulate mutation and recombination. Genetic programming repeats the evolutionary inspired step of fitness-based selection, crossover, and mutation until a predefined stop criterion is reached. In our case, we stop after a number of repeated evolutionary steps, which the GP literature often refers to as generations, but other schemes are also possible (Banzhaf et al., 1997).

Even though GP can be used in its original form to solve some problems, it is usually not possible to find a single motif able to separate between classes of DNA, RNA, or amino acid sequences. Also, single motifs produce binary outputs, whereas a dynamic range of scores is advantageous to assign probabilities to the predictions. Better classifiers can be constructed by including several motifs that by themselves may solve only parts of the problem. A simple solution would be to run GP several times, each of which would produce motifs with binary outcomes, and to define the predicted score as the average of the individual motifs' predictions (Breiman, 1996). Although this method will produce good results, better schemes exist. In this chapter, we will use boosting (Freund and Schapire, 1997; Meir and Rätsch, 2003) to create miRNA target classifiers.

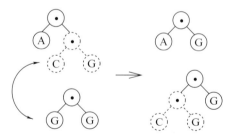

Figure 2.4 Query representation with cross-over example from genetic programming. The two queries on the left, ACG and GG, exchange parts of their trees and give rise to queries AG and CGG. This cross-over operation is complemented by mutations, which are spontaneous changes, such as deletions, substitutions, and insertions that affect single nodes or subtrees. Genetic programming evaluates the predictive performance of these queries, and attempts to breed better solutions as the algorithm propagate information from queries with merit from generation to generation.

Boosting is an iterative method that creates combined classifiers consisting of several basic classifiers, such as the GP motifs used here. Instead of creating GP motifs trained on identical datasets, boosting algorithms attempt to put weights on the samples so that some become more important. Typically, a bias is introduced so that the algorithm is pushed to learn to classify samples that were difficult for classifiers from previous iterations—the importance of certain samples are boosted. In our case, the final classifier becomes a set of weighted sequence motifs. Interestingly, boosting effectively renders GP a maximum margin classifier. Although each of the boosting iterations are still stochastic, the algorithm will converge to a maximum margin solution (Müller *et al.*, 2001). A mathematical connection between SVMs and boosted GP therefore exists.

4. miRNA Gene Prediction with Support Vector Machines

The most characteristic feature of miRNAs is that the primary transcript consists of an imperfect stem joined by a hairpin loop (Lagos-Quintana *et al.*, 2001; Lau *et al.*, 2001; Lee and Ambros, 2001). Some miRNAs are transcribed from their own promoter, others are excised from the introns of protein-coding genes, but they all fold into a hairpin structure (Bartel, 2004). The simplest miRNA gene prediction algorithm would therefore look for genomic hairpins in the genome. The problem is that the human genome encodes millions of potential hairpins. One thing is for certain—some hairpins may not be expressed at all, and even if some are, they may still belong to other classes of RNA. Nevertheless, because all miRNAs are hairpins, the first step of any miRNA gene prediction algorithm would be to identify hairpin candidates. The real test, however, will be the algorithm's ability to separate true miRNAs from other unrelated genomic hairpins.

We mentioned that the first miRNAs to be identified had been conserved throughout evolution. The first miRNA gene prediction algorithms took advantage of that and applied a filter to the initial list of genomic hairpins. Only hairpins conserved above a certain threshold were considered to be miRNA candidates (Grad *et al.*, 2003; Lai *et al.*, 2003; Lim *et al.*, 2003a,b). However, a class of nonconserved miRNAs was discovered (Bentwich *et al.*, 2005), which means that such a filter will hamper an algorithm's sensitivity for such genes. Some researchers may choose to apply a conservation filter to increase the positive predictive value before experimental validation, but this can be done to the output of any algorithm. Consequently, conservation filters will not be addressed throughout this chapter—neither for the SVM classifier for miRNA gene prediction, nor for the boosted GP classifier for miRNA target prediction.

The next sections follow the four machine learning steps previously outlined and present an SVM classifier for predicting miRNAs in *C. elegans*. We use the strategy of first creating a classifier that predicts the microprocessor processing site in a hairpin (Gregory *et al.*, 2004; Lee *et al.*, 2003) and then create a second classifier that uses these predictions to separate miRNAs from random genomic hairpins. This approach was developed to identify miRNAs in the human genome (Helvik *et al.*, 2006).

4.1. Deciding the input

Several groups have reported features that are important for miRNA biogenesis, which again—because these features can be assumed to be miRNA characteristics—can be used by a miRNA gene prediction algorithm. For example, elements outside the precursor's stem distal to the loop are important for pri-miRNA processing by the microprocessor complex (Han *et al.*, 2006). In contrast, others have reported that recognition of pri-miRNA depends on the distance of the cleavage site to the terminal loop (Zeng *et al.*, 2005). Furthermore, several groups have reported that efficient biogenesis downstream of the initial microprocessor cut depends on several sequence and structure features (Krol *et al.*, 2004; Ohler *et al.*, 2004; Zeng and Cullen, 2003, 2004). We reported that several characteristics are conserved among mammalian miRNA transcripts, suggesting that some features are indeed important for efficient biogenesis (Saetrom *et al.*, 2006).

Thus, as is often the case, a lot of information is available in the literature, but some of the evidence may be conflicting, and it is not easy to separate major from minor causalities. When creating a classifier, you want to use as much information as possible and to weight the various features according to their importance in anticipation that this will optimize the prediction performance. Note that we are not talking about extracting the most important features—such methods exist, however, and are called feature selection algorithms (Guyon and Elisseeff, 2003). Rather, it is only a question about classifier accuracy based on a set of input features. Thus, the SVM is a so-called black-box classifier, which means that building a biological model can be challenging.

We use several features related to miRNA primary and secondary structure, including the general and position-specific nucleotide composition in the miRNA precursor and its flanking regions, the precursor length and loop size, and the total position-specific occurrence of base pairs in the miRNA precursor and its flanking regions. See Helvik *et al.* (2006a) for a full list of features. We use RNAfold (Hofacker, 2003) with default parameters to predict RNA secondary structure and flanking regions of 50 nucleotides upstream and downstream of the precursor 5' and 3' ends.

When a decision has been made as to which features should be input to the SVM, these must be encoded in input vectors to the algorithm. Some features, such as the number of nucleotides from the predicted precursor and loop sizes, are numbers that can be appended to a vector. For other features, however, it is beneficial to retain the independence of various entities. For example, nucleotides in specific positions—adenine, thymine, cytosine, or guanine—could naïvely be encoded as four consecutive numbers. Because SVMs operate on Euclidian space vectors, however, it is better to encode the nucleotides as independent unity vectors in four dimensions, that is, as $(1,0,0,0)$, $(0,1,0,0)$, $(0,0,1,0)$, and $(0,0,0,1)$.

4.2. Assembling the training set

The current version of miRBase (release 8.2) contains 114 *C. elegans* miRNAs (Griffiths-Jones *et al.*, 2006), of which 112 have annotated mature sequences that have been mapped to the genome. These sequences will be our positive training set. The negative training set must, however, be tailored to each of the two classifiers.

For the classifier that predicts microprocessor processing sites—henceforth referred to as the microprocessor SVM—the negative training set will be the features found by considering all potential processing sites, except the true site given by the annotated location of the mature sequence. To limit the amount of negative examples, we will only consider sites that give precursor lengths between 50 and 80 nucleotides.

For the classifier that separates miRNAs from random genomic hairpins, the miRNA SVM, the negative training set will be random genomic hairpins. To construct this set, we first find all potential hairpins in the worm's genome and then select a random set of hairpins. We will select 3000 random hairpins to use as a negative training set and use the final classifier to predict miRNAs among the remaining hairpins extracted from the genome.

4.3. Training the miRNA gene predictor and estimating its performance

An SVM implementation is needed to train a classifier. A search for "SVM software" on an Internet search engine will reveal that there are several SVM packages available for free (see http://www.kernel-machines.org/index.html). For this work, we use the *gist* package (see http://microarray.cpmc.columbia.edu/gist/) with a radial basis kernel function and default parameter settings.

miRNAs can be grouped into families with similar mature sequences and secondary structures (Hertel *et al.*, 2006), and this similarity must be taken into account when the microprocessor SVM and miRNA SVM performances are estimated. As previously mentioned, mixing of similar examples in the training and test sets can give biased performance estimates.

Because of this potential bias, we use a variant of tenfold cross-validation where we place all miRNAs from the same family in the same fold. We use this approach for both the microprocessor and miRNA SVMs.

Both the microprocessor and miRNA SVM give a numeric score to each input. The scores comprise positive and negative values, and, even though it does not directly represent the probability that the input is the correct microprocessor processing site or is a true miRNA gene, the scores have a probabilistic interpretation. High positive scores mean that it is likely the input is the correct processing site or a true miRNA gene, and the converse is true for low negative scores. As previously described, a ROC curve shows how classifier sensitivity and specificity varies across all score thresholds and is normally a good indication of classifier performance. The ROC curve is a global performance measure that considers all the input independent, but for the microprocessor SVM, many inputs represent different potential processing sites in the same hairpin, and the main concern is whether the SVM gives the highest score to the correct processing site.

As Fig. 2.5A shows, the microprocessor SVM gives the highest score and thereby correctly predicts the processing site for almost 40% (37.4%) of the *C. elegans* miRNAs. Importantly, the predictions are not random, as the number of miRNAs with a wrong processing site scoring highest decreases with the number of sites scoring higher than the true site. Indeed, about 75% of the miRNAs have three or fewer false predictions scoring higher than the true site. If the predictions were random, all columns in the graph would have had approximately the same height.

For the miRNA SVM, we use the ROC curve to measure the classifier's performance, as each input represents a distinct genomic hairpin and therefore does not have the same interdependency as the input to the microprocessor SVM. Figure 2.5B shows that the miRNA SVM can indeed separate miRNAs from random genomic hairpins; the SVM finds 56, 75, and 85% of the miRNAs before the numbers of FP predictions reach 1, 5, and 10%.

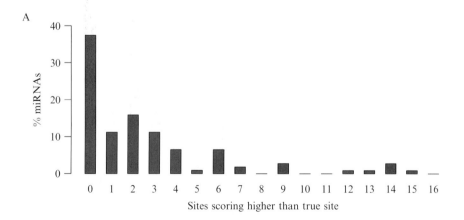

A

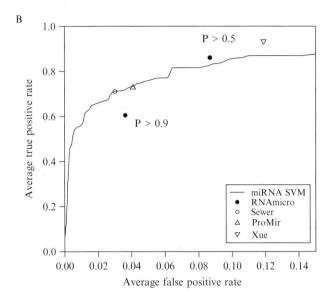

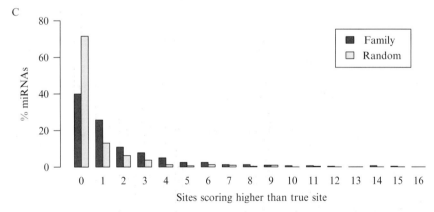

Figure 2.5 miRNA gene prediction results. (A) The number of sites that score higher than the true site is nonrandom. The leftmost bar implies that the top scoring site is the true site, which happens for almost 40% of the miRNAs. (B) The ROC curves show the relationship between average true positive rate (Se) and average false positive rate (1—Sp) for a range of score thresholds. Reported sensitivities and specificities are indicated for other algorithms, but recall that such measures are not directly comparable because the number of positive and negative examples in the data sets on which they were evaluated will affect the performance. (C) The difference between "family safe" and "random"cross-validations become evident in *Danio rerio*, where many miRNAs are highly similar and can be grouped in families. The performance estimate obtained from a "random" selection of folds in 10-fold cross-validation experiments is too high, as demonstrated by the significantly lower performance obtained when a "family safe" selection is used.

So how important are the modifications we made to the cross-validation routine to remove any potential bias by miRNAs from the same family being present in both the training and test set? To answer this question, we used standard 10-fold cross-validation and retrained and retested the microprocessor SVM. As it turns out, there is little difference in the performance estimates from the standard 10-fold cross-validation procedure and the "family-safe" cross-validation. Standard 10-fold cross-validation estimates that the microprocessor SVM predicts the correct processing site for 39.3% of the miRNAs, whereas the "family-safe" cross-validation estimate is 37.4%; both cross-validation methods estimate that 75.7% of the miRNAs have three or fewer false predictions scoring higher than the true site. The reason for this small difference between the two methods is that most verified *C. elegans* miRNAs are from different families— miRBase currently (release 8.2) groups the 113 *C. elegans* miRNAs into 102 distinct families.

In comparison, the 337 *Danio rerio* (zebrafish) miRNAs constitute only 92 distinct families, and we would therefore expect that standard cross-validation would overestimate the microprocessor SVM's performance on predicting microprocessor processing sites in zebrafish miRNAs. Indeed, standard cross-validation estimates that the microprocessor SVM predicts the correct processing site for 71.5% of the miRNAs, but the "family-safe" cross-validation estimate is 39.9% (see Fig. 2.5C), which is comparable to the *C. elegans* estimate. Thus, as seen with the zebrafish example, ensuring that the test set does not contain examples similar to those in the training set is important to get reliable classifier performance estimates.

4.4. Evaluating the miRNA gene predictor's performance

Many groups have developed methods that predict miRNA genes, but most methods rely on sequence or structure conservation to reduce the number of FP predictions (Bentwich, 2005; Berezikov *et al.*, 2006). Currently, four other methods do *ab initio* miRNA gene prediction without relying on sequence or structure conservation (Bentwich *et al.*, 2005; Helvik *et al.*, 2006; Nam *et al.*, 2005; Sewer *et al.*, 2005; Xue *et al.*, 2005); of these, three have estimated their predictive performance (Nam *et al.*, 2005; Sewer *et al.*, 2005; Xue *et al.*, 2005). To broaden the comparison, we also included a method that relies on sequence conservation to predict novel miRNAs (Hertel and Stadler, 2006). As Fig. 2.5B shows, this predictor has comparable performance to all these methods. Note, however, that all these estimates, except those for RNAmicro (Hertel and Stadler, 2006), are based on predictions of human miRNAs and may therefore not be accurate estimates of the methods' performance at predicting *C. elegans* miRNAs. For example, Sewer *et al.* (2005) and Xue *et al.* (2005) reported sensitivities of 43 and 86.4%, respectively, for *C. elegans* predictions, which is much lower than the 71 and 93.3% for *Homo sapiens* that is depicted in Fig. 2.5B. Furthermore, as we have previously mentioned, dataset characteristics, such as the number of true and false examples, may render the sensitivity and specificity performance measures unreliable.

5. miRNA Target Prediction with Boosted Genetic Programming

An miRNA target prediction classifier could be created along the lines of the previous section on miRNA gene prediction: Assemble a training set, identify important features from the literature, encode input vectors, and train and evaluate the SVMs performance using cross-validation. Indeed, others have used SVMs to predict miRNA target sites (Kim *et al.,* 2006). To demonstrate the use of a different machine learning algorithm, however, we will use boosted GP to create motif-based classifiers to predict miRNA target sites in 3' UTRs.

5.1. Deciding the input

Genetic programming requires that classifiers can be formalized in some language. Recall that a tree representation of the language facilitates evolutionary changes to the individual solutions that populate each generation (see Fig. 2.4). We use a search language that enables the use of a special-purpose search processor to evaluate the performance of each individual (Halaas *et al.,* 2004; Saetrom, 2004; Saetrom *et al.,* 2005)—GP is already computationally intensive and the boosted version even more so, but the search processor greatly accelerates the process. In principle, however, you can use any representation that computers can evaluate. The only catch is that the quality of the results may depend on your choice of solution language; for example, a language that cannot express the problem's solution will give poor results. Choosing the input language to GP therefore is a large part of deciding the algorithm input.

Here we follow the approach of Saetrom *et al.* (2005). The input to our classifier will be 30 nucleotide sequence windows representing verified target sites for *C. elegans* let-7 and lin-4 and *Drosophilia (D.) melanogaster* (fruit fly) miR-13a and bantam, and the corresponding miRNAs. Most of the target sites have much complementarity between both the 5' end 3' ends of the miRNAs, which indicates that these are mostly canonical or 3' complementary target sites (Brennecke *et al.,* 2005). We therefore use a solution language that consists of two recognition cores separated by a variable-length distance.

To understand the basis of our query language, consider the target site in Fig. 2.6. To find identical target sites for miR-13a in other mRNAs,

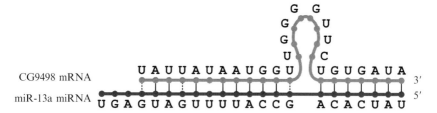

Figure 2.6   miRNA target site example; miR-13a from *D. melanogaster* targets CG9498 mRNA. Note the perfect complementarity between the miRNA's 5' end and the target, whereas some mismatches are allowed to the miRNA's 3' end.

we could simply search for the sequence UAUUA.AAUGGU........UGU-GAUA, where the wildcard . matches any nucleotide. This query will only find miR-13a target sites identical to the one in Fig. 2.6, however. To generalize this query to other miRNAs, we therefore replace the nucleotides with the corresponding positions in the miRNA to get the query $P_{19}P_{18}P_{17}P_{16}P_{15}.P_{13}P_{12}P_{11}P_{10}P_9P_8........P_7P_6P_5P_4P_3P_2P_1$.

In the final version of the query language, we have replaced the fixed-length wildcards that represent the target site's internal loop with variable-length wildcards. This way, the queries can match sites where the internal loop for example consists of between two and eight nucleotides. Furthermore, the recognition cores on each side of the loop wildcards ($p_{19}p_{18}P_{17}P_{16}P_{15}.P_{13}P_{12}P_{11}P_{10}P_9P_8$ and $P_7P_6P_5P_4P_3P_2P_1$ in the previous expression) can recognize target sites with some mismatches. More specifically, we can state that the recognition cores should match sequences that have an ungapped Smith Waterman distance below a specific threshold (Smith and Waterman, 1981). See Saetrom *et al.* (2005, 2006) for additional details.

5.2. Assembling the training set

As mentioned in the preceding section, the positive training set consists of verified target sites for *C. elegans* let-7 and lin-4 and *D. melanogaster* miR-13a and bantam, 36 in total (Boutla *et al.*, 2003; Brennecke *et al.*, 2003; Rajewsky and Socci, 2004). The negative training set will consist of 3000 random 30 nt 3' UTR sequence windows (Rajewsky and Socci, 2004; Saetrom *et al.*, 2005).

5.3. Training the miRNA target site predictor and estimating its performance

We use our custom GP and boosting software to train the miRNA target site predictor. This software was optimized to support the special-purpose search processor we use in our experiments (Halaas *et al.*, 2004), but the

Table 2.1 Comparison of "leave-one-miRNA-out" ROC-scores for various algorithms

Algorithm	References	let-7	lin-4	miR-13a	Bantam
		0.996	0.981	0.979	0.998
TargetBoost	(Saetrom *et al.*, 2005)	0.997	0.944	0.972	0.998
RNAhybrid	(Rehmsmeier *et al.*, 2004)	0.989	0.931	0.979	0.991
Nucleus	(Rajewsky and Socci, 2004)	0.988	0.962	0.928	0.998

The first row shows the result for our new predictor, which is a variant of TargetBoost with a slightly different query language. Compare the values to those of RNAhybrid and Nucleus.

software shares many of the features of freely available genetic programming packages (see http://www.geneticprogramming.com/GPpages/software.html and http://www.boosting.org/software.html for lists of genetic programming and boosting software).

As our training set only has verified target sites for four miRNAs, the target sites for the same miRNA will be present in both the training and test set for many of the 10 folds of a standard 10-fold cross-validation. To prevent any bias this may cause, we therefore use a cross-validation approach in which we use all the target sites from one miRNA as the test set and the remaining target sites as the training set. This gives four training and test sets. The first row of Table 2.1 summarizes the results of these "leave-one-miRNA-out" cross-validation experiments.

5.4. Evaluating the miRNA target site predictor's performance

We have previously shown that a version of our machine learning-based target site predictor with a slightly different query language has comparable and slightly better performance than two other target site prediction algorithms (Saetrom *et al.*, 2005). Table 2.1 compares the ROC-scores for these three algorithms to the current predictor and shows that our miRNA target site predictor is at least as good as the other algorithms. It is, however, important to note that the three algorithms will give slightly different predictions. To illustrate, our predictor does not tolerate GU wobble base pairs in the two recognition cores—RNAhybrid (Rehmsmeier *et al.*, 2004) does. Similarly, Nucleus (Rajewsky and Socci, 2004) requires a consecutive stretch of perfect base pairing (the nucleus), whereas our predictor tolerates mismatches in the recognition cores. Thus, depending on what part of the literature you believe, you could choose the algorithm that penalizes or tolerates GU wobbles and allows or denies mismatches in the miRNA

$5'$ end to predict the targets of your favorite miRNA (Didiano and Hobert, 2006; Doench and Sharp, 2004). The more likely road to success is, however, to use the predicted target sites common to all three methods.

6. SUMMARY

This chapter has presented a stepwise guide to machine learning and used this guide to create an SVM classifier for miRNA gene prediction and a GP classifier for miRNA target site prediction. Although the chapter focused on two specific machine learning algorithms, the stepwise guide is relevant for any learning approach. If you consider using a machine learning algorithm, or even creating your own classifier, to solve a particular problem, you should especially remember the importance of independent test sets for performance estimation. A classifier with biased performance estimates will look good on paper compared to a classifier with rigorous performance estimates, but which one would you use to guide your lab experiments? The answer to that question can only come from a classifier's long-term use as a research tool in the lab. Nevertheless, by getting good performance estimates and comparing the classifiers' performances and predictions, the classifier is more likely to be valuable in guiding validation experiments. In closing, this chapter has shown that knowledge about the problem domain is more important than machine learning expertise when designing an experiment. Consequently, we believe that machine learning is a valuable tool, and one that is possible to use for everyone, not just the experts in the machine learning field.

ACKNOWLEDGMENTS

The authors received support from the bioinformatics platform in the Norwegian Functional Genomics program (FUGE) and the Norwegian Research Council's Leiv Eriksson program. We thank O. R. Birkeland for valuable comments on the manuscript.

REFERENCES

Bagga, S., Bracht, J., et al. (2005). Regulation by let-7 and lin-4 miRNAs results in target mRNA degradation. Cell **122,** 553–563.

Bajic, V. B., Tan, S. L., et al. (2004). Promoter prediction analysis on the whole human genome. Nat. Biotechnol. **22,** 1467–1473.

Baldi, P., Brunak, S., et al. (2000). Assessing the accuracy of prediction algorithms for classification: An overview. Bioinformatics **16,** 412–424.

Banzhaf, W., Nordin, P., et al. (1997). "Genetic programming: An introduction: On the automatic evolution of computer programs and its applications." Morgan Kaufmann Publishers, San Francisco.

Bartel, D. P. (2004). MicroRNAs: Genomics, biogenesis, mechanism, and function. *Cell* **116,** 281–297.

Behlke, M. A. (2006). Progress towards *in vivo* use of siRNAs. *Mol. Ther.* **13**(4), 644–670.

Bentwich, I. (2005). Prediction and validation of microRNAs and their targets. *FEBS Lett.* **579,** 5904–5910.

Bentwich, I., Avniel, A., *et al.* (2005). Identification of hundreds of conserved and non-conserved human microRNAs. *Nat. Genet.* **37,** 766–770.

Berezikov, E., Cuppen, E., *et al.* (2006). Approaches to microRNA discovery. *Nat. Genet.* **38**(Suppl.), S2–S7.

Boutla, A., Delidakis, C., *et al.* (2003). Developmental defects by antisense-mediated inactivation of micro-RNAs 2 and 13 in *Drosophila* and the identification of putative target genes. *Nucl. Acids Res.* **31,** 4973–4980.

Breiman, L. (1996). Bagging predictors. *Machine Learning* **24**(2), 123–140.

Brennecke, J., Hipfner, D. R., *et al.* (2003). Bantam encodes a developmentally regulated microRNA that controls cell proliferation and regulates the proapoptotic gene hid in *Drosophila. Cell* **113,** 25–36.

Brennecke, J., Stark, A., *et al.* (2005). Principles of microRNA-target recognition. *PLoS Biol.* **3,** e85.

Burges, C. J. C. (1998). A tutorial on support vector machines for pattern recognition. *Data Mining Knowl. Disc.* **2,** 121–167.

D'Haeseleer, P. (2005). How does gene expression clustering work? *Nat. Biotechnol.* **23,** 1499–1501.

Dayhoff, J. E., and DeLeo, J. M. (2001). Artificial neural networks: Opening the black box. *Cancer* **91,** 1615–1635.

Didiano, D., and Hobert, O. (2006). Perfect seed pairing is not a generally reliable predictor for miRNA-target interactions. *Nat. Struct. Mol. Biol.* **13,** 849–851.

Doench, J. G., and Sharp, P. A. (2004). Specificity of microRNA target selection in translational repression. *Genes Dev.* **18,** 504–511.

Domingos, P., and Pazzani, M. (1997). On the optimality of the simple Bayesian classifier under zero-one loss. *Machine Learning* **29,** 103–130.

Esquela-Kerscher, A., and Slack, F. J. (2006). Oncomirs—microRNAs with a role in cancer. *Nat. Rev. Cancer* **6,** 259–269.

Freund, Y., and Schapire, R. E. (1997). A decision-theoretic generalization of on-line learning and an application to boosting. *J. Comput. System Sci.* **55,** 119–139.

Grad, Y., Aach, J., *et al.* (2003). Computational and experimental identification of C. elegans microRNAs. *Mol. Cell* **11,** 1253–1263.

Gregory, R. I., Yan, K. P., *et al.* (2004). The Microprocessor complex mediates the genesis of microRNAs. *Nature* **432**(7014), 235–240.

Griffiths-Jones, S., Grocock, R. J., *et al.* (2006). miRBase: MicroRNA sequences, targets and gene nomenclature. *Nucleic Acids Res.* **34**(Database issue), D140–D144.

Guyon, I., and Elisseeff, A. (2003). An introduction to variable and feature selection. *J. Machine Learning Res.* **3,** 1157–1182.

Halaas, A., Svingen, B., *et al.* (2004). A recursive MISD architecture for pattern matching. *Ieee Transact. Very Large Scale Integ. (Vlsi) Syst.* **12,** 727–734.

Han, J. J., Lee, Y., *et al.* (2006). Molecular basis for the recognition of primary microRNAs by the Drosha-DGCR8 complex. *Cell* **125,** 887–901.

Hebsgaard, S. M., Korning, P. G., *et al.* (1996). Splice site prediction in *Arabidopsis thaliana* pre-mRNA by combining local and global sequence information. *Nucleic Acids Res.* **24,** 3439–3452.

Helvik, S. A., and Snøve, O., Jr. (2007). Reliable prediction of Drosha processing sites improves microRNA gene prediction. *Bioinformatics* **23,** 142–149.

Hertel, J., Lindemeyer, M., *et al.* (2006). The expansion of the metazoan microRNA repertoire. *BMC Genomics* **7,** 25.

Hertel, J., and Stadler, P. F. (2006). Hairpins in a haystack: Recognizing microRNA precursors in comparative genomics data. *Bioinformatics* **22**, e197–e202.

Hofacker, I. L. (2003). Vienna RNA secondary structure server. *Nucleic Acids Res.* **31**, 3429–3431.

Jaakkola, T., Diekhans, M., *et al.* (2000). A discriminative framework for detecting remote protein homologies. *J. Comput. Biol.* **7**, 95–114.

Khan, J., Wei, J. S., *et al.* (2001). Classification and diagnostic prediction of cancers using gene expression profiling and artificial neural networks. *Nat. Med.* **7**, 673–679.

Kim, S. K., Nam, J. W., *et al.* (2006). miTarget: MicroRNA target-gene prediction using a Support Vector Machine. *BMC Bioinformatics* **7**, 411.

Koza, J. R., Bennett, H., *et al.* (1999). "Genetic Programming III: Darwinian Invention and Problem Solving." Morgan Kaufmann San Francisco.

Krol, J., Sobczak, K., *et al.* (2004). Structural features of microRNA (miRNA) precursors and their relevance to miRNA biogenesis and small interfering RNA/short hairpin RNA design. *J. Biol. Chem.* **279**, 42230–42239.

Lagos-Quintana, M., Rauhut, R., *et al.* (2001). Identification of novel genes coding for small expressed RNAs. *Science* **294**, 853–858.

Lai, E. C., Tomancak, P., *et al.* (2003). Computational identification of *Drosophila* micro-RNA genes. *Genome Biol.* **4**, R42.

Lau, N. C., Lim, L. P., *et al.* (2001). An abundant class of tiny RNAs with probable regulatory roles in *Caenorhabditis elegans*. *Science* **294**, 858–862.

Lee, R. C., and Ambros, V. (2001). An extensive class of small RNAs in *Caenorhabditis elegans*. *Science* **294**, 862–864.

Lee, Y., Ahn, C., *et al.* (2003). The nuclear RNase III Drosha initiates microRNA processing. *Nature* **425**, 415–419.

Lewis, B. P., Burge, C. B., *et al.* (2005). Conserved seed pairing, often flanked by adenosines, indicates that thousands of human genes are microRNA targets. *Cell* **120**, 15–20.

Lim, L. P., Glasner, M. E., *et al.* (2003a). Vertebrate microRNA genes. *Science* **299**, 1540.

Lim, L. P., Lau, N. C., *et al.* (2003b). The microRNAs of *Caenorhabditis elegans*. *Genes Dev.* **17**, 991–1008.

Lim, L. P., Lau, N. C., *et al.* (2005). Microarray analysis shows that some microRNAs downregulate large numbers of target mRNAs. *Nature* **433**, 769–773.

Lu, J., Getz, G., *et al.* (2005). MicroRNA expression profiles classify human cancers. *Nature* **435**, 834–838.

Martin, J. K., and Hirschberg, D. S. (1996). Small sample statistics for classification error rates I: Error rate measurements. *Technical Report*, 96–21, University of California, Irvine.

Meir, R., and Rätsch, G., eds. (2003). "An Introduction to Boosting and Leveraging. Advanced Lectures on Machine Learning." Springer-Verlag, Berlin.

Müller, K. R., Mika, S., *et al.* (2001). An introduction to kernel-based learning algorithms. *IEEE Trans. Neural Networks* **12**, 181–201.

Nam, J. W., Shin, K. R., *et al.* (2005). Human microRNA prediction through a probabilistic co-learning model of sequence and structure. *Nucleic Acids Res.* **33**, 3570–3581.

Ohler, U., Yekta, S., *et al.* (2004). Patterns of flanking sequence conservation and a characteristic upstream motif for microRNA gene identification. *RNA* **10**, 1309–1322.

Olsen, P. H., and Ambros, V. (1999). The lin-4 regulatory RNA controls developmental timing in *Caenorhabditis elegans* by blocking lin-14 protein synthesis after the initiation of translation. *Dev. Biol.* **216**, 671–680.

Pasquinelli, A. E., Reinhart, B. J., *et al.* (2000). Conservation of the sequence and temporal expression of let-7 heterochronic regulatory RNA. *Nature* **408**, 86–89.

Rajewsky, N., and Socci, N. D. (2004). Computational identification of microRNA targets. *Dev. Biol.* **267**, 529–535.

Rehmsmeier, M., Steffen, P., *et al.* (2004). Fast and effective prediction of microRNA/target duplexes. *RNA* **10,** 1507–1517.

Reinhart, B. J., Slack, F. J., *et al.* (2000). The 21-nucleotide let-7 RNA regulates developmental timing in *Caenorhabditis elegans*. *Nature* **403,** 901–906.

Saetrom, O., Snove, O., Jr., *et al.* (2005). Weighted sequence motifs as an improved seeding step in microRNA target prediction algorithms. *RNA* **11,** 995–1003.

Saetrom, P. (2004). Predicting the efficacy of short oligonucleotides in antisense and RNAi experiments with boosted genetic programming. *Bioinformatics* **20,** 3055–3063.

Saetrom, P., Snove, O., *et al.* (2006). Conserved MicroRNA characteristics in mammals. *Oligonucleotides* **16,** 115–144.

Safavian, S. R., and Landgrebe, D. (1991). A survey of decision tree classifier methodology. *IEEE Transact. Systems, Man Cybernet.* **21,** 660–674.

Salzberg, S. (1997). On comparing classifiers: Pitfalls to avoid and a recommended approach. *Data Mining Knowledge Disc.* **1,** 317–328.

Sewer, A., Paul, N., *et al.* (2005). Identification of clustered microRNAs using an ab initio prediction method. *BMC Bioinformatics* **6,** 267.

Smith, T. F., and Waterman, M. S. (1981). Identification of common molecular subsequences. *J. Mol. Biol.* **147,** 195–197.

Stone, M. (1974). Cross-validatory choice and assessment of statistical predictions. *J. Royal Stat. Soc. Series B (Methodological)* **36,** 111–147.

Vapnik, V. N. (1998). "Statistical Learning Theory." Wiley-Interscience, New York.

Wightman, B., Ha, I., *et al.* (1993). Posttranscriptional regulation of the heterochronic gene lin-14 by lin-4 mediates temporal pattern formation in *C. elegans. Cell* **75,** 855–862.

Wu, L., Fan, J., *et al.* (2006). MicroRNAs direct rapid deadenylation of mRNA. *Proc. Natl. Acad. Sci. USA* **103,** 4034–4039.

Xue, C., Li, F., *et al.* (2005). Classification of real and pseudo microRNA precursors using local structure-sequence features and support vector machine. *BMC Bioinformatics* **6,** 310.

Yekta, S., Shih, I. H., *et al.* (2004). MicroRNA-directed cleavage of HOXB8 mRNA. *Science* **304,** 594–596.

Zeng, Y., and Cullen, B. R. (2003). Sequence requirements for microRNA processing and function in human cells. *RNA* **9,** 112–123.

Zeng, Y., and Cullen, B. R. (2004). Structural requirements for pre-microRNA binding and nuclear export by Exportin 5. *Nucleic Acids Res.* **32,** 4776–4785.

Zeng, Y., Yi, R., *et al.* (2005). Recognition and cleavage of primary microRNA precursors by the nuclear processing enzyme Drosha. *EMBO J.* **24**(1), 138–148.

CHAPTER THREE

IDENTIFICATION OF VIRALLY ENCODED MICRORNAS

Sébastien Pfeffer

Contents

Abstract

RNA silencing is a widespread phenomenon that regulates gene expression at different levels. Small RNA molecules are at the core of all RNA silencing pathways and can be grouped in distinct families depending on their chemical properties and their mode of action. Among these small RNAs, microRNAs (miRNAs) represent one of the most extensively studied classes in animals. These tiny endogenous RNAs regulate gene expression posttranscriptionally by cleaving the targeted transcript or by interfering with its translation. miRNAs are found in plants and animals and have been identified in mammalian viruses. This indicates that mammalian viruses exploit the host RNA silencing machinery to produce miRNAs that have the potential to act both on the infected host genome and on the viral genome. The techniques used for identification of viral miRNAs mostly

Institut de Biologie Moléculaire des Plantes, CNRS, Strasbourg cedex, France

Methods in Enzymology, Volume 427
ISSN 0076-6879, DOI: 10.1016/S0076-6879(07)27003-X

parallel what is used for identifying cellular miRNAs. The use of prediction algorithms followed by validation has been successful. A more direct and nonbiased way of identifying viral miRNAs consists of cloning and sequencing small RNA libraries from virally infected cells or tissues.

1. INTRODUCTION

miRNAs are ~22 nt regulatory RNAs found in many eukaryotes that play major roles in very diverse biological processes (Bartel, 2004; Zamore and Haley, 2005). Only for human, there is currently a total of 474 miRNA genes listed in the miRBase sequence database (Griffiths-Jones et al., 2006), but the overall estimate is in the range of several thousands (Berezikov et al., 2005; Miranda et al., 2006), and each miRNA can potentially regulate the expression of tens of different genes. The biogenesis of these small RNAs is now partially elucidated, and shares common steps that are conserved from plants to mammals. First, a long primary transcript is transcribed by RNA polymerase II (Cai et al., 2004; Lee et al., 2004), and is recognized in the nucleus by the RNase III homolog Drosha that will cleave it in association with DGCR8/Pasha to release the stem-loop structured, ~70 nt long miRNA precursor (pre-miRNA) (Denli et al., 2004; Gregory et al., 2004; Landthaler et al., 2004; Lee et al., 2003; Zeng et al., 2005). The export factor Exportin 5 subsequently mediates shuttling of the pre-miRNA to the cytoplasm (Bohnsack et al., 2004; Lund et al., 2004; Yi et al., 2003), where a second RNase III–like enzyme, Dicer, cleaves it in association with TRBP, to excise the miRNA in the form of a small interfering RNA(siRNA)–like duplex (Chendrimada et al., 2005; Gregory et al., 2005; Haase et al., 2005; Hutvágner et al., 2001). The final step is asymmetric assembly of the mature miRNA into Argonaute protein containing silencing complexes. The mode of action of this loaded miRNP complex depends on the complementarity level of the miRNA with the targeted transcript. When there is perfect of near-perfect complementarity, the target is cleaved in a position corresponding to the middle of the miRNA. Argonaute 2 in mammals and Argonaute 1 in plants, have been shown to be responsible for this cleavage (Baumberger and Baulcombe, 2005; Liu et al., 2004; Meister et al., 2004). The cleaved RNA is then subjected to degradation by exonucleases (Orban and Izaurralde, 2005). The degradation mode of action is most common for plant miRNAs, whereas the preferred mode of action of animal miRNAs is translation inhibition after imperfect binding. Several hypotheses have emerged as to how this translation block occurs. For instance, the regulation step could be at the level of the transcript stability through deadenylation (Giraldez et al., 2006; Wu et al., 2006) or by sequestrating the targeted transcript to places where it will be inaccessible to the translation machinery, such as cytoplasmic processing bodies (Bhattacharyya *et al.*, 2006; Liu *et al.*, 2005; Pillai *et al.*, 2005). All these theories are not mutually exclusives, and it could be that different miRNAs act by different means.

The first viral miRNAs were discovered in the herpesvirus Epstein-Barr (EBV) (Pfeffer *et al.*, 2004). Although this finding was not completely unexpected, it illustrated a drastic difference regarding the role of RNA silencing in viral infections in mammals versus plants or insects. Indeed, it is now accepted that one major role of RNA silencing in these organisms is to protect them against viral infections. The fact that mammalian viral genomes are not recognized and degraded into siRNAs (Pfeffer *et al.*, 2005), but express miRNAs, that can potentially regulate host genes expression indicates that some mammalian viruses evolved toward pirating the cell's RNA silencing machinery rather than being targeted by it. Five miRNAs were originally identified in EBV by cloning and sequencing small RNA molecules from a latently infected cell line and were mapped to two loci on the viral genome. This initial finding triggered a massive search to look for miRNAs in other viruses. Viral miRNAs have now been found in Kaposi's sarcoma–associated herpesvirus, rhesus lymphocryptovirus, human cyto-megalovirus, murine herpesvirus 68, Marek's disease virus, SV40, and herpes simplex virus 1 (HSV1) (Burnside *et al.*, 2006; Cai *et al.*, 2005, 2006; Cui *et al.*, 2006; Dunn *et al.*, 2005; Grey *et al.*, 2005; Grundhoff *et al.*, 2006; Gupta *et al.*, 2006; Pfeffer *et al.*, 2004, 2005; Sullivan *et al.*, 2005). Viral miRNAs have the potential to regulate the expression of both viral (in cis) and cellular (in trans) genes. The *cis* mode of action is evidenced by the observation that the EBV miRNA miR-BART1 can target the transcript BALF5 laying exactly opposite to its location on the genome for degradation (Pfeffer *et al.*, 2004). A similar observation was made for SV40, which contains a miRNA opposite to the large T antigen transcript, resulting also in cleavage of this viral transcript (Sullivan *et al.*, 2005). The regulation of cellular genes by viral miRNAs has so far only been reported for HSV1, for which a miRNA, miR-LAT, has been shown to be involved in translational regulation of TGFβ and SMAD3, resulting in an antiapoptotic activity (Gupta *et al.*, 2006).

Numerous viral miRNAs remain to be identified, and the techniques available to this respect are similar to the ones used to identify cellular miRNAs, that is, bioinformatic predictions followed by validation or direct cloning and sequencing of small RNA libraries. However, it is worth noting that the initial search for new viral miRNAs raised the need for new predictions algorithms that do not rely on cross-species conservation. Indeed, viral miRNAs are not usually conserved with the host ones or with distantly related viruses. This chapter will focus on the small RNA cloning method from virally infected cells, and the reader is referred to publications in the literature regarding information on the prediction software tools available (Grundhoff *et al.*, 2006; Pfeffer *et al.*, 2005).

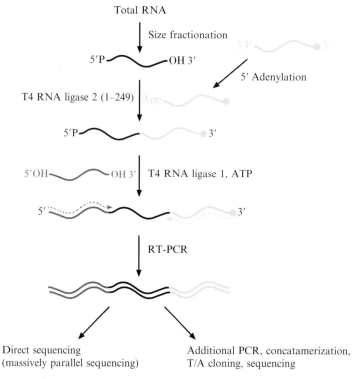

Figure 3.1 Schematic representation of the small RNA cloning protocol steps. (See color insert.)

The protocol described here, as diagrammed in Fig. 3.1, relies on the use of preadenylated adapter oligonucleotide that will be ligated to the 3′ of the small RNAs by the N terminal segment of T4 RNA ligase 2 (Rnl2 [1–249]). This truncated form of Rnl2 is impaired in adenylate transfer from ATP to the 5′ phosphate of RNA (the first step in the ligation process) (Ho *et al.*, 2004). This can be bypassed by the use of a preadenylated substrate that can be efficiently ligated to a 3′ hydroxyl group (second step of the ligation process) by Rnl2 (1–249). Therefore, circularization of the 5′ phosphorylated RNAs and their subsequent loss is minimized; in the next ligation step, only RNAs with a 5′ phosphate group (such as miRNAs) will be ligated to the 5′ adapter oligonucleotide by a classical T4 RNA ligase 1. Indirectly, this also results in an enrichment of the miRNAs fraction compared to classical cloning protocols that require dephosphorylation of the small RNA fraction prior to 3′ adapter ligation (Table 3.1). This feature can prove useful when cloning small RNA from cell lytically infected with viruses, in which the content in random degradation products of long RNAs can be quite important due to higher cell death.

Table 3.1 Comparison of the classic and optimized protocols

	Classic (%)	Optimized (%)
rRNA	51,85	1,59
tRNA	4,39	0,46
sn/snoRNA	1,12	0,07
Other ncRNA	0,17	0,40
miRNA	30,75	84,40
Other, not matched	11,71	13,09
Number clones	(1161)	(1513)

200 µg of total RNA from mouse NIH-3T3 cells was cloned following either the classic protocol using T4 RNA ligase 1 (Pfeffer *et al.*, 2003) or the protocol described in this chapter.

2. PURIFICATION OF RNL2 (1–249) AND ADENYLATION OF 3′ ADAPTER OLIGONUCLEOTIDE

Rnl2 (1–249) was expressed and purified from the plasmid pET16-Rnl2 (1–249) (a kind gift of S. Shuman) as described before (Ho and Shuman, 2002).

Adenylation of the 3′ adapter can be performed either chemically by synthesizing imidazole monophosphoadenylic acid (ImpA), which is then reacted with the phosphorylated oligonucleotide (Pfeffer *et al.*, 2003), or enzymatically. The chemical adenylation is a rather heavy procedure, and has the disadvantage of being very inefficient (10–20%). Wang and Silverman (2006) described an efficient way to adenylate enzymatically an oligonucleotide using T4 DNA ligase. The yield of 5′-AppRNA after gel purification can be as high as 80%. Briefly, 1 nmol of 5′ phosphorylated 3′ adapter (5′-TTTAACCGCGAATTCCAG-L-3′, L = blocking C7 amino group) and 3 nmol of a partially complementary oligonucleotide with a 10-nt overhang (5′-CTGGAATTCGCGGTTAAATATAGTGCAGT-3′) in 60 µl of 5 mM Tris (pH 7.5), 15 mM NaCl, and 0.1 mM EDTA were annealed by heating to 95° for 3 min and cooling on ice for 5 min. The solution was then brought to 100 µl total volume containing 40 mM MOPS (pH 7.0), 10 mM MgCl2, 10 mM DTT, 10 mM ATP, and 30 U T4 DNA ligase (MBI Fermentas) and incubated at 25° for 24 h. Samples were then loaded on a 20% denaturing acrylamide/bisacrylamide gel, and the band corresponding to the adenylated oligonucleotide was excised and eluted after visualization by ultraviolet (UV)-shadowing.

3. ISOLATION OF THE SMALL RNA FRACTION

Total RNA should be isolated by guanidinium thiocyanate-phenol-chloroform extraction (Chomczynski and Sacchi, 1987), which is commercially available as TRIzol (Invitrogen, CA) or Tri-reagent (Sigma-Aldrich). The required amount of starting material should ideally be in the range of 50 to 200 μg of total RNA dissolved in deionized formamide at a concentration of more than 2 mg/ml.

Prior to gel loading, RNA is diluted with 1 volume of a loading solution (50 mM EDTA, 99% formamide, 0.5 mg/ml bromophenol blue) and spiked with radiolabeled oligoribonucleotides of 19 and 24 nt in size (1 μl of a freshly made 20 nM stock) containing a Pme I restriction site (5′-CGUACGCGG*GUUUAAAC*GA-3′ and 5′-CGUACGCGGAAUA*GU-UUAAAC*UGU-3′; for all ribonucleotides, the Pme I restriction site is in italics). After denaturation for 30 sec at 95°, the RNA is loaded onto a 7 M Urea/15% acrylamide/bisacrylamide gel. The gel migrates for 30 to 45 min at 30 W using 0.5× TBE as a buffer until the bromophenol blue reaches the second third of the gel. The gel is dismounted, enveloped in a plastic wrap, and exposed to a phosphorimaging screen for 45 min. After printing out a 100%-scaled image, the gel is placed on the printout and a piece of gel defined by the RNA size markers is sliced out with a scalpel and cut into small pieces. Gel pieces are transferred to 1.5-ml siliconized tubes and weighted. Small RNAs are eluted from the gel after addition of 2 to 3 volumes of 0.3 M NaCl and constant agitation at 4° overnight. The supernatant is collected afterward, and RNA is precipitated at least 1 h at −20° after addition of 3 to 4 volumes of absolute ethanol.

4. LIGATION OF THE PURIFIED SMALL RNA TO ADENYLATED 3′ ADAPTER

After centrifugation for 10 min at maximum speed, the small RNA pellet is dissolved in 10 μl of water, 2 μl of 10× ligation buffer (0.1 M MgCl$_2$, 0.1 M 2-mercaptoethanol, 0.5 M Tris pH 7.6, and 1 mg/ml acetylated bovine serum albumin), and 6 μl of 50% DMSO. One μl of a 50 μM solution of the adenylated 3′ adapter is added, and the RNA is denatured for 30 sec at 90°. After cooling on ice, 0.1 μg of Rnl2 (1–249) is added, and the mixture is incubated at 37° for 1 h. The reaction is then stopped, and samples are separated on a denaturing 15% acrylamide/bisacrylamide gel. The ligation product is visualized after phosphorimaging of the gel for 45 min and is excised and eluted from the gel in 300 μl 0.3-M NaCl in siliconized tubes as described earlier. The RNA is subsequently precipitated by addition of 3 to 4 volumes of absolute ethanol.

5. LIGATION OF THE SMALL RNA-3′ ADAPTER TO THE 5′ ADAPTER

The small RNA pellet is collected by centrifugation and dissolved in 10 μl water, 2 μl of 10× ligation buffer supplemented with 2 mM ATP, 6 μl 50% DMSO, and 1 μl of 5′ adapter oligonucleotide

(5′-ACGGAATTCCTCACTrArArA-3′, r: ribonucleotide). After denaturation of the RNA 30 sec at 90°, 20 U of T4 RNA ligase (New England
Biolabs) are added, and the sample is incubated at 37° for 1 h. The reaction is
stopped by addition of 1 volume of formamide loading solution, and samples
are separated on a denaturing 15% acrylamide/bisacrylamide gel. The ligation
product is collected as in the previous step after phosphorimaging of the gel
and eluted overnight in 300 μl 0.3-M NaCl in siliconized tubes supplemented
with 1 μl of a 100 μM solution of the reverse transcription primer
(5′ GACTAGCTGGAATTCGCGGTTAAA 3′, all deoxyribonucleotides).
The next day, the RNA is precipitated by addition of 3 to 4 volumes of
absolute ethanol and storage at $-20°$.

6. REVERSE TRANSCRIPTION OF THE FINAL LIGATION PRODUCT

The small RNA ligated to the 3′ and 5′ adapters is collected by
centrifugation, dissolved in 5.6 μl of water, and denatured by heating the
tube for 30 sec at 90°. Then, 1.5 μl of 0.1 M DTT, 3 μl of 5× first-strand
buffer, and 4.2 μl of 2 mM dNTP are added, the reaction is incubated for
3 min at 50°, and 0.75 μl of reverse transcriptase (200 U/μl Superscript II,
RNase H [-] M-MLV reverse transcriptase, Invitrogen) are added. The
reaction is carried out at 42° for 30 min. The remaining RNA is hydrolyzed
by adding 40 μl of 150 mM KOH/20 mM Tris and incubating 10 min at
90°, and the solution is neutralized by addition of 40 μl of 150 mM Hcl to
obtain a pH between 7.0 and 9.5.

7. FIRST PCR AMPLIFICATION OF THE cDNA

To PCR–amplify the cDNA from previous step, 10 μl of 2 mM
dNTP, 10 μl of 10× PCR buffer, 1 μl of 100 μM 3′ primer (same as RT
primer), 1 μl of 5′ primer (5′ CAGCCAACGGAATTCCTCACTAAA 3′,
all deoxyribonucleotides), 67 μl of water, and 1 μl of Taq DNA polymerase
(5U/μl) are added to 10 μl of cDNA, and cycling is performed with the
following parameters: 45 sec at 94°, 85 sec at 50°, and 60 sec at 72° for 25
cycles. An aliquot of 5 μl of PCR is taken after 15, 20, and 25 cycles. The
PCR aliquots are visualized after loading on a 3% NuSieve agarose gel
(Cambrex) using a 25 bp DNA ladder (Invitrogen) as a size marker (the size
of the PCR product should be around 70 bp) to determine the number of
PCR cycles corresponding to the linear range of amplification. A new PCR
using the same conditions and the correct number of cycles is then performed. After amplification, add NaCl to a final concentration of 0.3 M and
extract the DNA once with 1 volume of phenol/chloroform and once with

1 volume of chloroform. The DNA is precipitated by the addition of 3 volumes of absolute ethanol.

8. Pme I Digestion of the PCR Product

To get rid of the cloned size markers, the PCR product is precipitated, the pellet is dissolved in 10 μl 10× NEB buffer 4, 88 μl water, and 2 μl Pme I and incubated at 37° for 2 h. The digested product is extracted once with phenol/chloroform after addition of NaCl to a final concentration of 0.3 M and once with chloroform and precipitated at least 1 h at −20° by addition of 3 volumes of absolute ethanol. To ensure the PCR to be used in the downstream reactions is clean, it is then gel-purified. To perform this step, the digested PCR product is pelleted and dissolved in 30 μl 1× PCR buffer, and is then loaded on a 3% NuSieve agarose gel. After migration, the band is visualized on a UV table, and cut out of the gel with a scalpel or razor blade. The gel slice is weighted, at least 1 volume of 0.4 M NaCl is added, and the tube is placed at 65° for 5 min. The DNA is then extracted once with 1 volume of phenol pH 8 previously heated to 65°, once with 1 volume of phenol/chloroform, and once with 1 volume of chloroform. After addition of 3 volumes of absolute ethanol, the PCR product is placed at −20° for at least 1 h for precipitation. After centrifugation, the pellet is dissolved in 50 μl of 1× PCR buffer.

9. Second PCR Amplification

To increase the quantity of DNA, and also to introduce Ban I restriction sites that will be used for concatamerization of the PCR products, an additional PCR is performed on the Pme I, digested first PCR product. One μl of the final product from the previous step is used in a PCR reaction containing 10 μl of 2 mM dNTP, 10 μl of 10× PCR buffer, 1 μl of 100 μM second PCR 5′ primer (5′ GAGCCAACAGGCACCGAATTCCTCACTAAA 3′, all deoxyribonucleotides), 1 μl of 100 μM second PCR 3′ primer (5′ GAC-TAGC-TTGGTGCCGAATTCGCGGTTAAA 3′, all deoxyribonucleotides), 76-μl water, and 1 μl Taq DNA polymerase. Fourteen PCR cycles are run as for the first PCR, and aliquots are taken after 8, 10, 12, and 14 cycles. The aliquots are examined on a 3% NuSieve gel as described earlier to determine the number of cycles corresponding to the linear range of amplification. The second PCR is then repeated in five times the volumes previously described. The reaction mixture is evenly divided between five PCR tubes. When the reaction is finished, NaCl is added to a final concentration of 0.3 M, and the PCR is extracted with 1 volume of phenol/chloroform followed by a 1-volume chloroform extraction. The DNA is precipitated with 2 volumes of absolute ethanol.

10. Ban I Digestion of the Second PCR Product

After centrifugation, the DNA pellet is dissolved in 1× Ban *I* buffer, and 10 *μ*l of 20 U/*μ*l Ban *I* are added (at this step, remember to keep 2 *μ*l of undigested product for further analysis). After 3 h incubation at 37°, the digested sample is analyzed on a 3% NuSieve gel side-by-side with the undigested aliquot to check for complete digestion. It can then be extracted with phenol/chloroform and chloroform as described previously.

11. Concatamerization of the Ban *I*-Digested DNA

After precipitation, the DNA is dissolved in 45 *μ*l of 1.1× T4 DNA ligation buffer, and 1.5 *μ*l of 100 *μM* second PCR 5′ primer and 1.5 *μ*l of 100 *μM* second PCR 3′ primer are added. After a 10 min incubation at 65°, 4 *μ*l of 2000 U/*μ*l T4 DNA ligase (New England Biolabs) are added and the reaction carried at 22° for 3 h. To check for complete concatamerization, 2 *μ*l of the reaction is analyzed on a 2% agarose gel with a 100-bp DNA ladder as a size marker. The concatamer should reach 1 kb and can be seen as a fuzzy ladder on the gel. The remainder of the reaction is then loaded on a 2% NuSieve agarose gel, a DNA band between 400 and 800 bp is excised with a scalpel, and the DNA is extracted from the gel as described previously.

12. End Tailing of Concatamers and Cloning into T/A Vector

After precipitation of the concatamers, the pellet is dissolved into 15 *μ*l of a PCR mix containing 1× PCR buffer, 0.2 m*M* dNTP, and 0.2 *μ*l Taq DNA polymerase and is incubated for 30 min at 72° for enzymatic 3′ tailing. This product can then directly be used in the TOPO cloning reaction as described by the manufacturer (Invitrogen). Typically, 4 *μ*l of the tailed product are reacted with 1 *μ*l of the pCR2.1 vector (Invitrogen) and 1 *μ*l of salt solution (1.2 *M* NaCl, 0.06 *M* MgCl$_2$) for 30 min at room temperature before being transformed into competent bacterial cells. After transformation, bacteria are plated on agar plates containing kanamycin (25 *μ*g/ml) and X-Gal.

13. Sequencing and Annotation of the Library

To screen the library, individual white colonies are resuspended into 20 μl of water in a 96-well plate. Ten μl of this bacterial suspension are then transferred to a second 96-well plate filled with 10 μl of a PCR mix composed of 2× PCR buffer, 100 μM dNTP, 3 mM MgCl2, 0.2 μM primer M13 (-20) forward primer (5′ GTAAAACGACGGCCAG 3′), 0.2 μM primer M13 reverse primer (5′ CAGGAAACAGCTATGAC 3′), and 0.075 U of Taq DNA polymerase, and cycling is performed using the following parameters: 94° for 90 sec, followed by 30 cycles of 94° for 30 sec, 57° for 30 sec, and 72° for 3 min and 30 sec. The PCR products are analyzed on a 2% standard agarose gel and submitted for sequencing using either primer M13 (-20) forward primer or M13 reverse primer. Using these PCR conditions has the advantage of depleting the reaction of dNTP and primers, and the product can thus be submitted directly for sequencing without the requirement of a PCR clean up.

Annotation of the library is the last step of the protocol. Small RNA sequences have to be extracted from the raw sequences by searching for adaptor sequences. Once a list of small RNA has been generated, these sequences have to be annotated by blasting them individually to various databases and to the genomes of both the organism from which the cells or tissue originated and of the virus infecting that organism. Typically, one will use a database containing a collection of all available miRNA sequences, such as miRBase (Griffiths-Jones *et al.,* 2006), a database compiling rRNA, tRNA, sno/snRNA sequences, and a database of mRNA sequences. All of these steps can be automated with the help of bioinformatics. Sequences that match to the viral genome and that were cloned multiple times have then to be examined to check whether they could be potential miRNAs. First a putative precursor sequence of 70 to 80 nt containing the cloned RNA must be extracted from the viral genome and folded using programs such as mfold (Zuker, 2003). If the fold-back structure obtained resembles that of a pre-miRNA, it then has to be shown that it can be detected by other techniques such as Northern blot before it can be affirmed that it represents a true viral miRNA.

14. Concluding Remarks

Viral miRNAs were discovered recently, and many more remain to be identified. Although the biggest challenge ahead of us will clearly be the identification of cellular targets of viral miRNAs, important information will be generated as the list of these sequences expand. The generalization of

massively parallel sequencing approaches (Margulies *et al.*, 2005) will facilitate the collection of large number of sequences. Because it seems each virus
family has evolved its own set of miRNAs, we can learn a lot from
comparing viral miRNAs to cellular miRNAs in various aspects such as
biogenesis, expression, and target recognition. So far, all viral miRNAs that
have been described were isolated from cell lines; it will be of interest also to
see how viral miRNAs accumulate in the course of a natural infection. In
that sense, cloning small RNA libraries from patients samples will prove
very informative.

REFERENCES

Bartel, D. P. (2004). MicroRNAs: Genomics, biogenesis, mechanism, and function. *Cell*
   116, 281–297.
Baumberger, N., and Baulcombe, D. C. (2005). Arabidopsis ARGONAUTE1 is an RNA
   Slicer that selectively recruits microRNAs and short interfering RNAs. *Proc. Natl. Acad.
   Sci. USA* **102,** 11928–11933.
Berezikov, E., Guryev, V., van de Belt, J., Wienholds, E., Plasterk, R. H., and Cuppen, E.
   (2005). Phylogenetic shadowing and computational identification of human microRNA
   genes. *Cell* **120,** 21–24.
Bhattacharyya, S. N., Habermacher, R., Martine, U., Closs, E. I., and Filipowicz, W.
   (2006). Relief of microRNA-mediated translational repression in human cells subjected
   to stress. *Cell* **125,** 1111–1124.
Bohnsack, M. T., Czaplinski, K., and Gorlich, D. (2004). Exportin 5 is a RanGTP-
   dependent dsRNA-binding protein that mediates nuclear export of pre-miRNAs.
   RNA **10,** 185–191.
Burnside, J., Bernberg, E., Anderson, A., Lu, C., Meyers, B. C., Green, P. J., Jain, N.,
   Isaacs, G., and Morgan, R. W. (2006). Marek's disease virus encodes microRNAs that
   map to meq and the latency-associated transcript. *J. Virol.* **80,** 8778–8786.
Cai, X., Hagedorn, C. H., and Cullen, B. R. (2004). Human microRNAs are processed
   from capped, polyadenylated transcripts that can also function as mRNAs. *RNA* **10,**
   1957–1966.
Cai, X., Lu, S., Zhang, Z., Gonzalez, C. M., Damania, B., and Cullen, B. R. (2005).
   Kaposi's sarcoma-associated herpesvirus expresses an array of viral microRNAs in latently
   infected cells. *Proc. Natl. Acad. Sci. USA* **102,** 5570–5575.
Cai, X., Schafer, A., Lu, S., Bilello, J. P., Desrosiers, R. C., Edwards, R., Raab-Traub, N.,
   and Cullen, B. R. (2006). Epstein-Barr virus microRNAs are evolutionarily conserved
   and differentially expressed. *PLoS Pathog.* **2,** e23.
Chendrimada, T. P., Gregory, R. I., Kumaraswamy, E., Norman, J., Cooch, N.,
   Nishikura, K., and Shiekhattar, R. (2005). TRBP recruits the Dicer complex to Ago2
   for microRNA processing and gene silencing. *Nature* **436,** 740–744.
Chomczynski, P., and Sacchi, N. (1987). Single-step method of RNA isolation by acid
   guanidinium thiocyanate-phenol-chloroform extraction. *Anal. Biochem.* **162,** 156–159.
Cui, C., Griffiths, A., Li, G., Silva, L. M., Kramer, M. F., Gaasterland, T., Wang, X. J., and
   Coen, D. M. (2006). Prediction and identification of herpes simplex virus 1-encoded
   microRNAs. *J. Virol.* **80,** 5499–5508.
Denli, A. M., Tops, B. B., Plasterk, R. H., Ketting, R. F., and Hannon, G. J. (2004).
   Processing of primary microRNAs by the microprocessor complex. *Nature* **432,**
   231–235.

Dunn, W., Trang, P., Zhong, Q., Yang, E., van Belle, C., and Liu, F. (2005). Human cytomegalovirus expresses novel microRNAs during productive viral infection. *Cell Microbiol.* **7,** 1684–1695.

Giraldez, A. J., Mishima, Y., Rihel, J., Grocock, R. J., Van Dongen, S., Inoue, K., Enright, A. J., and Schier, A. F. (2006). Zebrafish MiR-430 promotes deadenylation and clearance of maternal mRNAs. *Science* **312,** 75–79.

Gregory, R. I., Chendrimada, T. P., Cooch, N., and Shiekhattar, R. (2005). Human RISC couples microRNA biogenesis and posttranscriptional gene silencing. *Cell* **123,** 631–640.

Gregory, R. I., Yan, K. P., Amuthan, G., Chendrimada, T., Doratotaj, B., Cooch, N., and Shiekhattar, R. (2004). The Microprocessor complex mediates the genesis of micro-RNAs. *Nature* **432,** 235–240.

Grey, F., Antoniewicz, A., Allen, E., Saugstad, J., McShea, A., Carrington, J. C., and Nelson, J. (2005). Identification and characterization of human cytomegalovirus-encoded microRNAs. *J. Virol.* **79,** 12095–12099.

Griffiths-Jones, S., Grocock, R. J., van Dongen, S., Bateman, A., and Enright, A. J. (2006). miRBase: MicroRNA sequences, targets and gene nomenclature. *Nucleic Acids Res.* **34,** D140–D144.

Grundhoff, A., Sullivan, C. S., and Ganem, D. (2006). A combined computational and microarray-based approach identifies novel microRNAs encoded by human gamma-herpesviruses. *RNA* **12,** 733–750.

Gupta, A., Gartner, J. J., Sethupathy, P., Hatzigeorgiou, A. G., and Fraser, N. W. (2006). Anti-apoptotic function of a microRNA encoded by the HSV-1 latency-associated transcript. *Nature* **442,** 82–85.

Haase, A. D., Jaskiewicz, L., Zhang, H., Laine, S., Sack, R., Gatignol, A., and Filipowicz, W. (2005). TRBP, a regulator of cellular PKR and HIV-1 virus expression, interacts with Dicer and functions in RNA silencing. *EMBO Rep.* **6,** 961–967.

Ho, C. K., and Shuman, S. (2002). Bacteriophage T4 RNA ligase 2 (gp24.1) exemplifies a family of RNA ligases found in all phylogenetic domains. *Proc. Natl. Acad. Sci. USA* **99,** 12709–12714.

Ho, C. K., Wang, L. K., Lima, C. D., and Shuman, S. (2004). Structure and mechanism of RNA ligase. *Structure (Camb.)* **12,** 327–339.

Hutvágner, G., McLachlan, J., Pasquinelli, A. E., Balint, E., Tuschl, T., and Zamore, P. D. (2001). A cellular function for the RNA-interference enzyme Dicer in the maturation of the let-7 small temporal RNA. *Science* **293,** 834–838.

Landthaler, M., Yalcin, A., and Tuschl, T. (2004). The human DiGeorge syndrome critical region gene 8 and its *D. melanogaster* homolog are required for miRNA biogenesis. *Curr. Biol.* **14,** 2162–2167.

Lee, Y., Ahn, C., Han, J., Choi, H., Kim, J., Yim, J., Lee, J., Provost, P., Radmark, O., Kim, S., and Kim, V. N. (2003). The nuclear RNase III Drosha initiates microRNA processing. *Nature* **425,** 415–419.

Lee, Y., Kim, M., Han, J., Yeom, K. H., Lee, S., Baek, S. H., and Kim, V. N. (2004). MicroRNA genes are transcribed by RNA polymerase II. *EMBO J.* **23,** 4051–4060.

Liu, J., Carmell, M. A., Rivas, F. V., Marsden, C. G., Thomson, J. M., Song, J. J., Hammond, S. M., Joshua-Tor, L., and Hannon, G. J. (2004). Argonaute2 is the catalytic engine of mammalian RNAi. *Science* **305,** 1437–1441.

Liu, J., Valencia-Sanchez, M. A., Hannon, G. J., and Parker, R. (2005). MicroRNA-dependent localization of targeted mRNAs to mammalian P-bodies. *Nat. Cell Biol.* **7,** 719–723.

Lund, E., Guttinger, S., Calado, A., Dahlberg, J. E., and Kutay, U. (2004). Nuclear export of microRNA precursors. *Science* **303,** 95–98.

Margulies, M., Egholm, M., Altman, W. E., Attiya, S., Bader, J. S., Bemben, L. A., Berka, J., Braverman, M. S., Chen, Y. J., Chen, Z., Dewell, S. B., Du, L., *et al.* (2005). Genome sequencing in microfabricated high-density picolitre reactors. *Nature* **437,** 376–380.

Meister, G., Landthaler, M., Patkaniowska, A., Dorsett, Y., Teng, G., and Tuschl, T. (2004). Human Argonaute 2 mediates RNA cleavage targeted by miRNAs and siRNAs. *Mol. Cell* **15,** 185–197.

Miranda, K. C., Huynh, T., Tay, Y., Ang, Y. S., Tam, W. L., Thomson, A. M., Lim, B., and Rigoutsos, I. (2006). A pattern-based method for the identification of microRNA binding sites and their corresponding heteroduplexes. *Cell* **126,** 1203–1217.

Orban, T. I., and Izaurralde, E. (2005). Decay of mRNAs targeted by RISC requires XRN1, the Ski complex, and the exosome. *RNA* **11,** 459–469.

Pfeffer, S., Lagos-Quintana, M., and Tuschl, T. (2003). Cloning of small RNA molecules. *In* "Current Protocols in Molecular Biology" (F. M. Ausubel, R. Brent, R. E. Kingston, D. D. Moore, J. G. Seidmann, J. A. Smith, and K. Struhl, eds.), pp. 26.4.1–26.4.18. Wiley Interscience, New York.

Pfeffer, S., Sewer, A., Lagos-Quintana, M., Sheridan, R., Sander, C., Grasser, F. A., van Dyk, L. F., Ho, C. K., Shuman, S., Chien, M., Russo, J. J., Ju, J., *et al.* (2005). Identification of microRNAs of the herpesvirus family. *Nat. Methods* **2,** 269–276.

Pfeffer, S., Zavolan, M., Grasser, F. A., Chien, M., Russo, J. J., Ju, J., John, B., Enright, A. J., Marks, D., Sander, C., and Tuschl, T. (2004). Identification of virus-encoded micro-RNAs. *Science* **304,** 734–736.

Pillai, R. S., Bhattacharyya, S. N., Artus, C. G., Zoller, T., Cougot, N., Basyuk, E., Bertrand, E., and Filipowicz, W. (2005). Inhibition of translational initiation by let-7 microRNA in human cells. *Science* **309,** 1573–1576.

Sullivan, C. S., Grundhoff, A. T., Tevethia, S., Pipas, J. M., and Ganem, D. (2005). SV40-encoded microRNAs regulate viral gene expression and reduce susceptibility to cytotoxic T cells. *Nature* **435,** 682–686.

Wang, Y., and Silverman, S. K. (2006). Efficient RNA 5′-adenylation by T4 DNA ligase to facilitate practical applications. *RNA* **12,** 1142–1146.

Wu, L., Fan, J., and Belasco, J. G. (2006). MicroRNAs direct rapid deadenylation of mRNA. *Proc. Natl. Acad. Sci. USA* **103,** 4034–4039.

Yi, R., Qin, Y., Macara, I. G., and Cullen, B. R. (2003). Exportin-5 mediates the nuclear export of pre-microRNAs and short hairpin RNAs. *Genes Dev.* **17,** 3011–3016.

Zamore, P. D., and Haley, B. (2005). Ribo-gnome: The big world of small RNAs. *Science* **309,** 1519–1524.

Zeng, Y., Yi, R., and Cullen, B. R. (2005). Recognition and cleavage of primary micro-RNA precursors by the nuclear processing enzyme Drosha. *EMBO J.* **24,** 138–148.

Zuker, M. (2003). Mfold web server for nucleic acid folding and hybridization prediction. *Nucleic Acids Res.* **31,** 3406–3415.

CHAPTER FOUR

COMPUTATIONAL METHODS FOR MICRORNA TARGET PREDICTION

Yuka Watanabe,*,† Masaru Tomita,*,†,‡ *and* Akio Kanai*,†,‡

Contents

Abstract

The discovery of microRNAs (miRNAs) has introduced a new paradigm into gene regulatory systems. Large numbers of miRNAs have been identified in a wide range of species, and most of them are known to downregulate translation of messenger RNAs (mRNAs) via imperfect binding of the miRNA to a specific site or sites in the 3′ untranslated region (UTR) of the mRNA. Identification of genes targeted by miRNAs is widely believed to be an important step toward understanding the role of miRNAs in gene regulatory networks. As part of the effort to understand interactions between miRNAs and their targets, computational algorithms have been developed based on observed rules for features such as the degree of hybridization between the two RNA molecules. These *in silico* approaches provide important tools for miRNA target detection, and together with experimental validation, help to reveal regulated targets of miRNAs. Here, we summarize the knowledge that has been accumulated about the principles of target recognition by miRNAs and the currently available computational methodologies for prediction of miRNA target genes.

* Institute for Advanced Biosciences, Keio University, Tsuruoka, Japan
† Systems Biology Program, Graduate School of Media and Governance, Keio University, Fujisawa, Japan
‡ Faculty of Environment and Information Studies, Keio University, Fujisawa, Japan

Methods in Enzymology, Volume 427
ISSN 0076-6879, DOI: 10.1016/S0076-6879(07)27004-1

1. INTRODUCTION

MicroRNAs (miRNAs) comprise a small class of noncoding RNA (ncRNA) genes identified in the genomes of various species. Discovery of novel gene regulatory systems under the control of small ncRNAs has had a significant impact on molecular biology. These RNAs are known to recognize and bind to partially complementary sites in 3′ untranslated regions (3′UTRs) of mRNAs to regulate translation of the mRNA (Bartel, 2004). Over 4000 miRNAs have been identified from a number of different species (Griffiths-Jones, 2004). In addition, it is thought that functional characterization of miRNAs will depend heavily on identification of their specific target mRNAs. However, experimental studies have touched on only a handful of the possible ranges of function of miRNAs, and numerous bioinformatic methods were developed to allow high-throughput prediction of miRNA target genes (Bentwich, 2005; Brown and Sanseau, 2005; Rajewsky, 2006; Yoon and De Micheli, 2006).

Computational prediction of miRNA target sites consists of four main steps: (1) extraction of rules related to formation of miRNA–mRNA duplexes; (2) incorporation of those rules in computational algorithms; (3) prediction of novel miRNA target sites using those algorithms; and (4) validation of the results, and thus the algorithm itself, using computational and experimental approaches. In the case of animal miRNA targets, the fact that miRNAs are very short and miRNA–mRNA duplexes are not entirely complementary made the elucidation of hybridization patterns rather challenging. Computational and experimental approaches have revealed that not only the binding pattern but also the relationship between the targeting miRNAs play an important role within miRNA target recognition (Bentwich, 2005; Brown and Sanseau, 2005; Rajewsky, 2006; Yoon and De Micheli, 2006). There are numerous useful resources and software tools available for analysis of miRNA targets, as summarized in Tables 4.1 and 4.2. Results derived using these computational algorithms have been validated biologically, and feedback from validation results have greatly improved performance of *in silico* miRNA target prediction algorithms.

Here, we will focus on the efforts made to predict miRNA target genes and understand the biology of miRNAs. We summarize principles of miRNA target recognition, available resources for computational prediction of miRNA target sites, and validation strategies for computational prediction. This comprehensive survey on analysis of miRNA targets is expected to provide an overall view of the knowledge that has been accumulated in the field.

Table 4.1 Databases of microRNAs and their targets

Name of the database	URL	Main features	Reference(s)
mirBase	http://microrna.sanger.ac.uk/	miRNA sequences, annotations, and computationally predicted targets	Griffiths-Jones et al., 2004, 2006
Argonaute	http://www.ma.uni-heidelberg.de/apps/zmf/argonaute/interface/	Detailed information about known miRNAs and their targets	Shahi et al., 2006
miRNAMap	http://mirnamap.mbc.nctu.edu.tw/	Known and computationally predicted miRNAs and their targets	Hsu et al., 2006
Tarbase	http://www.diana.pcbi.upenn.edu/tarbase.html	Experimentally verified miRNA targets	Sethupathy et al., 2006
Arabidopsis Small RNA Project Database	http://asrp.cgrb.oregonstate.edu/	Arabidopsis miRNA sequences and corresponding target genes in addition to other small RNAs	Gustafson et al., 2005

Table 4.2 Computational algorithms for microRNA target prediction

Name of the software	URL or availability	Supported organism(s)	Reference(s)
TargetScan, TargetScanS	http://genes.mit.edu/targetscan/	Vertebrates	Lewis et al., 2003, 2005
miRanda	http://www.microrna.org/	Flies, vertebrates	Enright et al., 2003, John et al., 2004
DIANA-microT	http://diana.pcbi.upenn.edu/DIANA-microT/	Vertebrates	Kiriakidou et al., 2004
RNAhybrid	http://bibiserv.techfak.uni-bielefeld.de/rnahybrid/	Flies	Rehmsmeier et al., 2004
GUUGle	http://bibiserv.techfak.uni-bielefeld.de/guugle/	Flies	Gerlach et al., 2006
PicTar	http://pictar.bio.nyu.edu/	Nematodes, flies, vertebrates	Grun et al., 2005, Krek et al., 2005, Lall et al., 2006
MicroInspector	http://mirna.imbb.forth.gr/microinspector/	Any	Rusinov et al., 2005
MovingTargets	Available by request on DVD	Flies	Burgler et al., 2005
FastCompare	http://tavazoielab.princeton.edu/mirnas/	Nematodes, flies	Chan et al., 2005
miRU	http://bioinfo3.noble.org/miRNA/miRU.htm	Plants	Zhang 2005
TargetBoost	https://demo1.interagon.com/demo/	Nematodes, flies	Saetrom et al., 2006
rna22	http://cbcsrv.watson.ibm.com/rna22.html	Nematodes, flies, vertebrates	Miranda et al., 2006
miTarget	http://cbit.snu.ac.kr/~miTarget/	Any	Kim et al., 2006

2. PRINCIPLES OF miRNA TARGET RECOGNITION

It has been challenging to predict target genes for miRNAs in animals because of the complexity of miRNA target recognition. This complexity includes the fact that miRNAs are short and that miRNA–mRNA duplexes are not entirely complementary to one another. To develop computational algorithms that identify miRNA target genes, empirical evidence is examined carefully, and principles of miRNA target recognition are extracted. For example, base pairing between miRNAs and their targets can be analyzed and checked for features, such as stable binding at 5' end of the miRNA. Next, thermodynamic analysis of miRNA–mRNA duplexes is performed via calculation of the free energy of duplex formation and evaluation of the thermodynamics of binding. Then, cross-species sequence comparison is used to ask whether the target sequence has been evolutionarily conserved between related species. Finally, the number of target sites for the miRNA is counted, as suggested by previous studies that mRNAs are likely to be regulated by miRNA binding at more than one target site. These criteria can be applied stepwise, such as in the pipeline shown in Fig. 4.1 (Bentwich, 2005; Brown and Sanseau, 2005; Rajewsky, 2006; Yoon and De Micheli, 2006).

There is a specific base pairing pattern within miRNA–mRNA duplexes (Fig. 4.2), and it has been suggested that there is some importance to the pattern. In particular, the 5' region of the miRNA is perfectly complementary to the corresponding sequence of target mRNA in most cases, and is generally very well conserved among paralogous miRNAs. Analysis of patterns of complementarity has been performed for vertebrates (Lewis et al., 2003), humans (Kiriakidou et al., 2004), flies (Brennecke et al., 2005; Stark et al., 2003), and worms (Lall et al., 2006; Watanabe et al., 2006). The 7 to 8 base pair sequence starting from either the first or second base of the 5' end of the miRNA is referred to as the "seed" region of the miRNA–mRNA duplex. G:U wobble base pairing is thought to be kept at a minimum within seed regions, as weak pairing reduces silencing efficiency (Doench and Sharp, 2004). At the same time, base pairing at the 3' region of miRNA is thought to be weaker and less important (Brennecke et al., 2005; Kiriakidou et al., 2004; Stark et al., 2003; Vella et al., 2004; Watanabe et al., 2006). However, some reports indicate that stronger binding at the 3' region can compensate for weaker binding in the 5' seed (Brennecke et al., 2005; Doench and Sharp, 2004). Also, there is at least one case in which the target gene is regulated independently of base pairing in the seed region (Didiano and Hobert, 2006). In contrast to the strong binding at seed region, weaker binding is observed at the central region of the miRNA–mRNA duplex (Brennecke et al., 2005; Kiriakidou et al., 2004; Stark et al., 2003; Vella et al., 2004; Watanabe et al., 2006). Mismatches in the central region of miRNA–mRNA duplex form a bulge structure, which may be significant for regulation of target gene

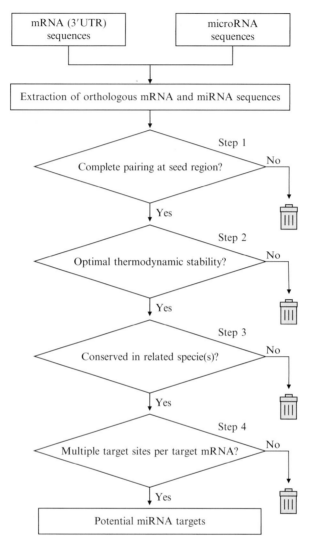

Figure 4.1 A pipeline for miRNA target prediction. The main steps in identifying miRNA target genes are shown. When miRNA and mRNA (3'UTR) sequences are provided as input data sets, similar data sets from related species are constructed using data on putative orthologs. After preparation of the data sets, miRNA binding sites are identified by determining the base pairing pattern of miRNAs and mRNAs according to the complementarity within specific regions (Step 1); determining the strength of the resulting miRNA–mRNA duplex by calculation of the free energy (Step 2); comparative sequence analysis (Step 3); and checking for the presence of multiple target sites per transcript (Step 4).

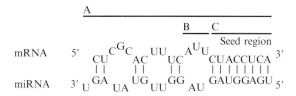

Figure 4.2 A typical pattern of base pairing between miRNAs and target mRNAs. Typically, the miRNA binds to a specific site or sites within the 3′UTR region of the mRNA sequence. According to thermodynamic analysis, some degree of complex formation occurs along the entire miRNA–mRNA duplexed region (A). Base pairing is particularly weak in the central region (B) and particularly strong at the 5′ end (seed region) of the miRNA (C). These aspects are commonly used to identify putative novel binding sites. Base pairing between *let-7* miRNA and *hbl-1* mRNA in *C. elegans* is shown as an example (Lin *et al.*, 2003).

expression. Taken together, these facts suggest the possibility that the mechanism of miRNA recognition differs between species or cellular conditions and also that commonalities between recognition mechanisms are not fully understood.

In most of miRNA target prediction algorithms, the thermodynamic properties of miRNA–mRNA duplex formation are analyzed by calculation of free energy (ΔG), which is considered an important aspect for evaluation. The Vienna package, which can be used to estimate free energy and secondary structure, is the most commonly used software tool for thermodynamic analysis (Wuchty *et al.*, 1999). With this tool, the free energy threshold is calculated based on both specificity and sensitivity, and as such, the values are likely to differ for different organisms. For example, free energy thresholds were set to -14 kcal/mol for *Drosophila (D.) melanogaster* (Enright *et al.*, 2003), -17 kcal/mol for *Homo sapiens* (John *et al.*, 2006), and -20 kcal/mol for *Caenorhabditis (C.) elegans,* (Watanabe *et al.*, 2006). However, it is very difficult to determine the appropriate thresholds of free energy, as data sets of known miRNA–mRNA duplexes are very limited, and a lower free energy (stable binding) does not always result in reliable prediction of miRNA target genes. Therefore, it is necessary to take other characteristics into account. Toward that end, one report demonstrates a method to effectively predict miRNA targets without using thermodynamics and by using conservation analysis instead (Lewis *et al.*, 2005).

The degree of sequence conservation is another criterion commonly used to filter miRNA targets from genome sequence. Many of the target prediction algorithms identify orthologous 3′UTR sequences and then check whether the potential miRNA target site is conserved in other species, such as via comparison of relatively closely related species, including *C. elegans* and *Caenorhabditis (C.) briggsae* or *D. melanogaster* and *Drosophila pseudoobscura* (Brennecke *et al.*, 2005; Burgler and Macdonald, 2005; Enright *et al.*, 2003; Kiriakidou *et al.*, 2004; Rajewsky and Socci, 2004; Rehmsmeier, 2006; Stark *et al.*, 2003; Watanabe *et al.*, 2006). Generally, $3'$ UTR sequences of different species are aligned using software tools such as BLASTn (Waterman and Eggert, 1987) or AVID (Bray *et al.*, 2003; Couronne *et al.*, 2003), and conservation of the sequence and/or position within $3'$UTR sequence are checked. Moreover, the University of California Santa Cruz (UCSC) genome database (Karolchik *et al.*, 2003) provides genome-wide alignment for a wide range of species and also is commonly used for conservation analysis.

Another way of incorporating phylogeny into the extraction scheme is by comparing sequences from a wide range of species (Chan *et al.*, 2005; Grun *et al.*, 2005; Krek *et al.*, 2005; Lall *et al.*, 2006; Wang and Wang, 2006). This is rather challenging, as noncoding RNAs and $3'$UTR regions are generally less conserved than protein coding regions, and thus, there is a risk of a substantial rate of false negatives (i.e., cases in which true miRNA targets are not recognized as such). At the same time, it has also been suggested that incorporating a wide range of evolutionary conservation into the target extraction pipeline improves the signal-to-noise ratio of the extracted miRNA target candidates and allows for high accuracy of prediction of miRNA target genes (Rajewsky, 2006). Therefore, users must determine appropriate methods and thresholds according to what they wish to gain from a given analysis.

A number of studies have shown that more than one miRNA can potentially bind to a single targeted gene and that together, multiple miRNAs may cooperatively control the expression of target genes (Enright *et al.*, 2003; John *et al.*, 2004; Watanabe *et al.*, 2006). Computational evidence suggests cooperation in miRNA target interactions occurs in a manner different than what can be expected by chance interactions (Stark *et al.*, 2005). Also, some experimental evidence indicates that the degree of miRNA target regulation differs according to the number of miRNAs that bind to a target gene (Krek *et al.*, 2005; Vella *et al.*, 2004). This feature of multiple target sites may contribute to fine adjustment of target gene expression or may act as a back-up for essential gene regulation, possibilities that remain to be proven.

One report suggests that prediction tools can be improved by incorporating the folded structure of the mRNA into the prediction algorithm (Robins *et al.*, 2005). The way that this group took secondary structure into account was based on their hypothesis that single-stranded miRNAs can only reach potential target sites when stretches of the target mRNA do not form base pairs with another part of mRNA. Thus, they first checked whether the mRNA is folded in a way that the site of interest is base paired with another part of mRNA by predicting the secondary structure of the $3'$UTR sequence. By taking the structure of mRNA into account, the algorithm was reportedly able to locate target sites without relying on evolutionary conservation.

The principles previously described are useful for prediction of miRNA targets in animals. For plants, miRNA target prediction is rather straightforward, as plant miRNAs are believed to base pair to their targets with perfect or nearly perfect antisense complementarity (Llave *et al.*, 2002). Attempts have been made to detect miRNA target genes in plants, primarily for *Arabidopsis (A.) thaliana* (Bonnet *et al.*, 2004; Li and Zhang, 2005; Rhoades *et al.*, 2002; Wang *et al.*, 2004). Target search in *A. thaliana* is carried out in two main steps: (1) detection of mRNA regions that are complementary or nearly complementary to miRNA sequences; and (2) identification of orthologous sites of miRNA complementarity in related species. Because plant miRNA bind to their targets with perfect or nearly perfect base pairing, some have speculated that plant miRNAs may act similarly to small interfering RNAs (siRNAs), contributing to cleavage of target mRNA (Rhoades *et al.*, 2002).

3. RESOURCES FOR ANALYSIS OF miRNA TARGET GENES

As our understanding of miRNAs and their target genes has improved, some useful databases have been constructed and serve as a registry of this information (Table 4.1). In 2004, a comprehensive database of miRNA sequences was constructed, and it has been updated as new miRNAs have been identified (Griffiths-Jones, 2004; Griffiths-Jones *et al.*, 2006). The authors of this database also proposed a uniform definition for miRNAs (Ambros *et al.*, 2003). As increasing numbers of miRNAs have been revealed, several databases that house information about miRNA targets have also been constructed (Gustafson *et al.*, 2005; Hsu *et al.*, 2006; Sethupathy *et al.*, 2006; Shahi *et al.*, 2006). In this section, resources which can be very useful to initiate study of miRNA targeted genes are described.

The miRNA registry, or miRBase, is a well-known and widely used database of miRNA sequences (Griffiths-Jones, 2004; Griffiths-Jones *et al.*, 2006). This database was initially developed to assign uniform names to miRNAs. A web interface (http://microrna.sanger.ac.uk/) allows users to search and browse for miRNAs in a wide range of species, and provides a downloadable flat file containing miRNA sequence information via the FTP site. The database is also open to submission of newly identified miRNAs. Since its initial release, the database has been updated to include novel miRNA sequences. The current release 8.2 contains 4039 entries from 45 different species. The miRBase has been expanded to include not only miRNA sequence data but also information about the potential genomic targets of miRNAs. The miRBase Target database provides predicted miRNA target genes for various species, relying on the miRBase Sequences database as a data source for miRNA sequences and on the miRanda software as a miRNA target prediction algorithm (John *et al.*, 2004; Stark *et al.*, 2003).

The Argonaute database provides comprehensive and precise information about miRNAs and their target genes (Shahi *et al.*, 2006). For example, information about expression patterns, proposed or known functions, and target genes are provided via a web interface (http://www.ma.uni-heidelberg.de/apps/zmf/argonaute/interface/). A comprehensive list of miRNA families and proteins involved in processing of miRNA are also provided. This database is currently composed of 893 miRNAs from *H. sapiens, Mus (M.) musculus,* and *Rattus (R.) norvegicus.*

The miRNAMap database includes known and putative miRNAs and their known and putative target genes, and is available at http://mirnamap. mbc.nctu.edu.tw/ (Hsu *et al.*, 2006). This database integrates experimental evidence pertaining to miRNAs and their targets culled from existing databases and published literature. In addition, putative miRNAs are predicted using RNAz software (Washietl *et al.*, 2005), and putative miRNA targets are predicted using miRanda software (John *et al.*, 2004; Stark *et al.*, 2003). Information is available at miRNAMap for four mammalian species, *H. sapiens, M. musculus, R. norvegicus,* and *Canis familiaris.*

TarBase is a manually curated collection of experimentally supported miRNA targets and can be accessed at http://www.diana.pcbi.upenn.edu/tarbase.html. This database covers a wide range of species, from plants to human, and provides information on a total of 763 target sites at 570 target genes (Fig. 4.3). Also, information about false targets (predicted miRNA-target interaction with negative experimental results) can be found in this database.

Most of the aforementioned databases have been constructed using information about animal miRNA targets. By contrast, the *Arabidopsis* Small RNA Project (ASRP) database provides detailed data about *A. thaliana* miRNAs and their targets (Gustafson *et al.*, 2005). This database was originally constructed as a repository for small RNA sequences cloned by the ASRP but now also serves as a comprehensive database of *Arabidopsis* miRNAs and their experimentally validated and computationally predicted target genes. This database is available through a web interface at http://asrp.cgrb.oregonstate.edu/.

4. Software Useful for miRNA Target Prediction

As an increasing number of miRNAs from various species are identified, demand for prediction of new miRNA target sites using optimal miRNA and mRNA sequences has increased. Software with different

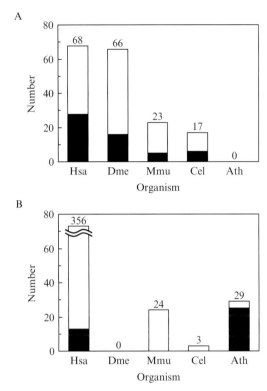

Figure 4.3 miRNA target sites supported by experimental evidence. The number of miRNA-mRNA pairs identified either experimentally (open bars) or informatically and validated experimentally (solid bars) for *Homo sapiens* (Hsa), *Mus musculus* (Mmu), *Drosophila melanogaster* (Dme), *Caenorhabditis elegans* (Cel), and *Arabidopsis thaliana* (Ath). **A,** miRNAs that translationally repress target mRNAs. Original counts were obtained from TarBase version 2 (Sethupathy *et al.*, 2006). **B,** miRNAs that cleave target mRNAs.

characteristics has been developed and applied for prediction of a large number of target genes (Brown and Sanseau, 2005; Yoon and De Micheli, 2006). In this section, we describe 11 software tools useful for miRNA target prediction and prediction results derived from use of those software tools. The algorithms we describe are listed in Table 4.2, including their sources, suitable species, and references.

TargetScan is an algorithm developed by Lewis *et al.* (2003) for prediction of miRNA targets in vertebrates. This software combines thermodynamics-based miRNA–mRNA duplex prediction and comparative sequence analysis. Lewis *et al.* observed perfect Watson-Crick complementarity at bases 2 to 8 in the 5′ end of miRNAs (the seed sequence). They incorporate this feature into their prediction algorithm by searching for seed matches against mRNA sequences and extending the seed matches to predict the remaining extent of miRNA–mRNA binding. The thermodynamic properties of binding between the putative miRNA target and the extended seed sequence is calculated using the RNAfold program (Hofacker, 2003). Phylogenic analysis was carried out using genomic sequence from human, mouse, rat, and pufferfish. Eleven of 15 predicted targets generated using this software were experimentally validated. Moreover, the false-positive rate was estimated at 31% for mammalian miRNA targets, and the software was used to predict 451 potential miRNA targets. Although functions of the predicted target genes encompassed a broad range, the group was enriched for genes involved in transcriptional regulation.

Later, the authors improved the TargetScan algorithm and proposed TargetScanS, which requires a shorter seed match (six nucleotides), is independent of thermodynamic stability or multiple target sites, and requires the presence of conserved adenosine around the seed sequence (Lewis *et al.*, 2005). The authors also added two more species to their cross-species conservation analysis, dog and chicken. These changes reduced the estimated false-positive rate of the algorithm to 22% in mammals. The algorithm successfully predicted all of the known miRNA-target interactions and in total, results in prediction of over 5300 human genes as potential targets of miRNAs. Thus, these analyses suggest that over one third of human genes are controlled by miRNAs.

The miRanda software was initially designed to predict miRNA target genes in *D. melanogaster* (Enright *et al.*, 2003; John *et al.*, 2004). This algorithm consists of three basic phases: (1) identification of a sequence that may be bound to miRNA; (2) calculation of the free energy for predicted miRNA–mRNA duplexes; and (3) detection of evolutionary conservation among *D. melanogaster, D. pseudoobscura,* and *Anopheles gambiae.* This method correctly identified 9 of 10 published miRNA-target interactions. The false-positive rate was estimated at 24%. As previously reported, the set of predicted target gene functions was enriched for transcription factors, reaffirming the possible importance of miRNA in development. This algorithm was also applied to prediction of human miRNA targets (Enright *et al.*, 2003; John *et al.*, 2004). About 2000 putative human miRNA target genes were identified, suggesting that 10% or more of human genes are regulated by miRNAs.

Kiriakidou *et al.* (2004) searched for miRNA binding properties using an experimental approach and incorporated the properties they identified into a computational algorithm, DIANA-microT. In contrast to previously discussed approaches, with this approach the authors addressed the necessity for a central bulge and strong binding at 3′ end of miRNA when 5′ seed pairing is rather weak. Also, in contrast of the previous works, this method uncovers predominant miRNA targets that contain only single target sites.

This algorithm successfully identified all of the documented *C. elegans* miRNA-target pairs. Moreover, seven predicted mammalian miRNA target genes were validated experimentally.

Rehmsmeier *et al.* (2004) presented the RNAhybrid program, which is an extension of classical RNA secondary structure prediction software tools, such as RNAfold (Hofacker, 2003) and Mfold (Mathews *et al.*, 1999). This software searches for energetically optimal binding sites for a small RNA within a large RNA sequence. This algorithm was applied to miRNA target search in *D. melanogaster* using a six nucleotide seed match starting from the second base of the 5' end of the miRNA. A currently released utility program called GUUGle (Gerlach and Giegerich, 2006), which can locate potential helical regions in RNA sequence, may be combined with RNAhybrid to improve the speed of search for miRNA target genes.

Cross-species comparisons provide powerful criteria for identifying miRNA target genes, and PicTar software fully relies on comparative data from several species to identify common targets for miRNAs (Krek *et al.*, 2005). Moreover, PicTar computes the maximum likelihood that a given sequence is bound by one or more miRNAs. Target genes are first predicted using common criteria, such as optimal binding free energy, and are then tested statistically using genome-wide alignment of eight vertebrate genomes to filter out false positives. The false-positive rate for PicTar has been estimated to be about 30%, and known miRNA target sequences were identified correctly using this software. Krek *et al.* used this algorithm to predict vertebrate miRNA targets and suggested that on average, approximately 200 transcripts are regulated by a single miRNA. They were able to experimentally validate 7 of 13 mouse miRNA target candidates.

Using the PicTar algorithm with cross-species comparison of seven *Drosophila* species, Grun *et al.* (2005) were able to predict miRNA targets in *D. melanogaster*. The results suggest that on average, 54 genes are regulated by a given miRNA and that *D. melanogaster* miRNAs regulate expression of target genes coordinately. PicTar was also applied to genome-wide search of miRNA targets in *C. elegans* (Lall *et al.*, 2006). For this study, the researchers analyzed cross-species conservation for three nematodes. The results suggested at least 10% of *C. elegans* genes are predicted miRNA targets. The authors also speculate that miRNAs regulate biological processes by targeting genes that are functionally related to one another.

The web tool MicroInspector was developed for identification of miRNA binding sites within targets from a variety of species (Rusinov *et al.*, 2005). The simple web server allows a user to assign mRNA sequences, miRNA dataset (categorized by organism), hybridization temperature, and a free energy cut-off for calculation of possible miRNA target sites. This web tool successfully located all of the known miRNA–mRNA interactions.

MovingTargets software can be used to predict a set of miRNA targets that fulfill a set of biological constraints (Burgler and Macdonald, 2005). This algorithm applies five biological constraints: (1) the number of target sites in one mRNA; (2) thermodynamic strength of miRNA–mRNA binding; (3) the number of consecutive seed pairings at the $5'$ end of the miRNA; (4) the total number of $5'$ nucleotides in the miRNA involved in target base pairing; and (5) the number of G:U wobbles at the $5'$ region of the miRNA. Applying this software to *D. melanogaster* revealed a set of 83 high-likelihood miRNA targets. Three of these candidates were tested and all three could be experimentally verified.

Chan *et al.* (2005) proposed a rather unique way of detecting miRNA target sites. In contrast to previous approaches, the approach used by this group relies on searching for target sequences before using miRNA sequences. Highly conserved mRNA motifs were detected via network-level conservation using FastCompare software, and these were considered potential miRNA target sites. Next, the researchers searched for preexisting miRNAs with complementarity to this set of potential targets. If existing miRNAs could not be identified, the researchers went on to search for novel miRNAs in genomic sequence. In this way, target sites were searched within *D. melanogaster* and *D. pseudoobscura* and *C. elegans* and *C. briggsae* genome pairs, and the results suggested there is likely a large number of miRNAs that have not yet been identified and validated.

As recognition mechanism seems to differ between animal and plant miRNAs, software for detecting miRNA targets from animals cannot be used to detect plant miRNAs. miRU is a web server developed specifically for plant miRNA target gene prediction (Zhang, 2005). Because plant miRNAs recognize their target mRNAs via perfect or nearly perfect base pairing, this algorithm searches for potential complementary target sites that approach perfect complementarity. This software is capable of detecting any plant miRNA target site and can search for evolutionarily conserved miRNA target sites when genome sequence is available for an appropriate comparison species. True positives and false positives can also be estimated based on the characteristics observed within predicted miRNA–mRNA duplexes.

The TargetBoost algorithm relies on training with a set of known miRNA targets, which are used to create weighted sequence motifs that capture characteristics common to validated miRNA–mRNA binding sites (Saetrom *et al.*, 2005). The authors indicated that the weighted sequence motif approach used in TargetBoost incorporates both duplex stability and complementarity. Using a set of 36 experimentally verified target sites as the training set, the authors of TargetBoost were able to predict potential miRNA target sites for genes involved in *D. melanogaster* body patterning.

Miranda *et al.* (2006) introduced a pattern-based method for prediction of miRNA target sites. This software, rna22, first searches for putative miRNA target sites and then identifies possible targeting miRNAs. The rna22 program was used to successfully predict miRNA target sites without the use of cross-species validation. Subsequently, 226 predicted target genes were tested in a luciferase reporter gene assay, and for 168 of them, miRNA-dependent repression was observed. They suggest that some miR-NAs may regulate as many as a few thousand target genes and up to 74 to 92% of all transcripts in some species.

A support vector machine (SVM) classifier has also been applied to miRNA target prediction in the form of the miTarget software tool (Kim *et al.*, 2006). miTarget incorporates structural, thermodynamic, and positional information as inputs for SVM-based analysis. This method was reported to predict a biologically relevant set of miRNA targets with higher performance than previously published tools according to the receiver operating characteristic (ROC) curve analysis. The authors also indicate that base pairing at positions four, five, and six in the 5' region of the miRNA is particularly important.

5. Original Strategies for Prediction of miRNA Target Genes

A number of reports present specific pipelines for extracting miRNA target sites. Those methods are not incorporated as software, but their methodologies and prediction results give a good idea of how and what kinds of genes are regulated by miRNAs. Also, in some of these reports, miRNA target prediction is performed for species such as plants and viruses, whose miRNA target recognition mechanisms may differ greatly from that of animals.

Stark *et al.* (2003) first characterized hybridization patterns of known miRNA–mRNA duplexes in *D. melanogaster* and used the information to extract target sites with conserved 8mer nucleotides at the 5' end of the miRNAs in addition to taking into account evolutionary conservation with a closely related species, *D. pseudoobscura*. This group used the MFold package (Mathews *et al.*, 1999) to calculate the thermodynamic stability of miRNA–mRNA hybridization. Known miRNA target genes scored high with their approach. In addition, the importance of multiple miRNA target sites per target gene was also suggested by their results. Six of the predicted miRNA target candidates were tested and validated.

Brennecke *et al.* (2003) used a reporter gene assay to extract hybridization characteristics of miRNAs and their targets for further computational analysis. From their experimental analysis, these researchers found that target sites can be grouped into two types: target sites with high complementarity at the 5' end of the miRNA, and target sites with 3' base pairing and a low degree of complementarity at the 5' end. These characteristics were incorporated in a computational algorithm and applied to genome-wide extraction of *D. melanogaster* miRNA targets. The analysis provides evidence that an average miRNA has approximately 100 target sites and that a large number of cellular genes are under the control of miRNAs.

Systematic identification of human miRNA target genes was carried out by combining an *in silico* miRNA target prediction method with microarray analysis (Wang and Wang, 2006). This group first predicted miRNA targets in the human genome using an algorithm that combines relevant parameters for miRNA target recognition and calculates a score according to the different weights assigned to each of these parameters. Predicted target genes for miR-124 were systematically verified using microarray analysis of target genes in the context of overexpression of miR-124. From the analysis, significant down regulation of cell cycle–related genes was observed.

To elucidate target sites within the *C. elegans* genome, a target extraction pipeline was constructed according to the characteristics of known miRNA–mRNA duplexes of *C. elegans* (Watanabe *et al.*, 2006). Extraction of miRNA target candidates was performed in three steps: overall binding of miRNA and mRNA was predicted using the RNAhybrid software (Rehmsmeier *et al.*, 2004); hybridization patterns were selected and conservation between *C. elegans,* and a related nematode, *C. briggsae,* was taken into account; and finally, the number of target sites within a single target gene was determined. Free energy was calculated for every dinucleotide of the miRNA–mRNA duplex to analyze the hybridization pattern more precisely than that achieved via reliance on analysis of simple complementarity. This method predicted 687 potential target genes in *C. elegans,* many of which are classified as genes involved in development.

Most of the miRNA target prediction algorithms have been constructed for finding miRNA targets in animals. This is likely to be the results of the observation that base pairing and other attributes of miRNA-target interactions are complicated and not uniform, such that different rules can be applied within a species, and in addition, different rules may apply to different species. In contrast, miRNA–target interactions in plants appear to more heavily rely on base pairing. Although there are fewer algorithms available, a number of plant miRNA targets have been predicted as they can rely on straightforward base pairing searches and conservation analyses. For example, Rhoades *et al.* (2002) predicted regulatory targets for 14 *A. thaliana* miRNAs, and extracted 49 candidate miRNA targets. Thirty-four of these miRNA target candidates are members of transcription factor gene families, suggesting the possible contribution of miRNA to transcription control mechanisms, similar to what was observed in animals. Also, Jones-Rhoades and Bartel (2004) developed a comparative genomic algorithm that can systematically identify miRNAs and their targets. Nineteen novel miRNA target candidates, conserved in both *A. thaliana* and

Oryza sativa, are biologically relevant. Targets predicted using this algorithm also tend to be genes involved in development.

Virus-encoded miRNA has been identified (Pfeffer *et al.*, 2004), and possible targets for viral miRNAs were predicted using the miRanda algorithm (Enright *et al.*, 2003). Another report suggested the existence of a novel type of miRNA which functions in mammalian antiviral immunity (Lecellier *et al.*, 2005), indicating that there may be a complicated miRNA-mediated interaction between viruses and their hosts. Computational methods for identification of viral miRNA targets have yet to be developed, likely due to a lack of experimentally analyzed examples of interaction between viral miRNAs and their targets. However, there is a report of computational search for cellular targets carried out using predicted HIV-I encoded miRNAs (Bennasser *et al.*, 2004). Additionally, a search for human miRNA targets in HIV-1 genes has also been reported (Hariharan *et al.*, 2005). In that study, the authors used four well-established target-finding software tools in their analysis, and some known features (e.g., seed matching and cross-species comparison) were also taken into account in their extraction of possible target sites. Crucial HIV-1 genes, including the *nef* gene, were predicted to be possible miRNA target genes. There remains much to be learned about the mechanisms of miRNA regulation as it relates to viruses, and it is very difficult to carry out target prediction based solely on the existing knowledge. Nonetheless, computational target prediction may be able to contribute greatly to the understanding of miRNA-mediated interactions between viruses and their hosts in the future.

6. VALIDATION OF COMPUTATIONAL PREDICTIONS

Validation of computational prediction algorithms and predicted miRNA targets is crucial for understanding the biological significance of prediction results. Moreover, feedback of validation results should be useful for the further analysis of predicted targets and evaluation of computational algorithms. There are two main strategies used for validation of miRNA target prediction software: evaluation of known miRNA target genes, and calculation of signal-to-noise ratio or false-positive rates using negative control tests (i.e., evaluation using artificial miRNA-like sequences).

Most miRNA prediction algorithms report the significance of their method by showing how well the algorithm performs in terms of correctly identifying known miRNA–mRNA interactions. Thus, a limitation of this method of reporting the accuracy of predictions would be the lack of an available positive data set. Only a small number of miRNA target interactions have been experimentally analyzed at this point (see Fig. 4.3); therefore, it is difficult to determine statistical significance. Moreover, training sets used in the development of algorithms are usually extracted from this small dataset of validated targets, making the dataset that can subsequently be used for evaluation even smaller. Saetrom *et al.* (2005) used a method similar to cross-validation to help overcome these problems. With their method, accuracy of the algorithm was determined using a dataset of known miRNA–mRNA duplexes.

Another strategy commonly used for evaluation of miRNA target prediction is comparing prediction results calculated from real and artificial miRNA data inputs. Artificial miRNAs are shuffled sequences that resemble real miRNAs in terms of features such as base composition, sequence length, and frequency of appearance in the genome. By comparing the prediction results derived from real and artificial miRNAs and determining how the results differ, a signal-to-noise ratio or false-positive rate can be estimated.

Experimental validation of miRNA target interactions is crucial to detecting novel miRNA targets, as computational methods are not perfect, and there is a risk of false-positive prediction. Although experimental validation of miRNA target genes is challenging compared to computational validation, there are more and more examples of miRNA target genes from various species that were identified using combined computational and biological approaches (see Fig. 4.3). Experimental validation is performed against two types of predicted miRNA targets that have different regulatory mechanisms: translational repression of target mRNAs (see Fig. 4.3A) and cleavage of target mRNAs (see Fig. 4.3B). Methods such as reporter-gene assays (Burgler and Macdonald, 2005; Kiriakidou *et al.*, 2004; Krek *et al.*, 2005; Miranda *et al.*, 2006; Robins *et al.*, 2005), gene mutation (Brennecke *et al.*, 2005; Stark *et al.*, 2003, 2005), and 5′ RACE (Jones-Rhoades and Bartel, 2004; Wang *et al.*, 2004) have commonly been used for experimental validation of computationally predicated miRNA targets that are translationally regulated. In addition, microarray analysis provides a powerful and high-throughput method for observing cleaved target mRNAs (Wang and Wang, 2006). For example, this experimental approach was used to identify a large number of human miRNA targets that appear to be cleaved by miRNAs (see Fig. 4.3B; Lim *et al.*, 2005).

7. Concluding Remarks

Prediction and identification of miRNA target genes should be the first steps toward understanding the biology of miRNAs. Computational and experimental approaches have revealed numerous characteristics of miRNA-target interactions and have led to effective prediction of target sites. The accumulated knowledge about miRNA target recognition has revealed the principles of miRNA–mRNA duplex formation, such as strong base pairing at the 5' seed region of the miRNA. However, there are some exceptions to these generalized rules, and it is also true that target selection mechanisms differ among species. To improve the accuracy of miRNA target prediction, phylogenetic analysis has been applied and has been used to successfully predict a number of target genes. Moreover, experimental validation, including feedback of validation data to computational tools, has also played an important role in identification of computationally predicted miRNA target genes.

Combining experimental analysis with computational approaches should be very effective in developing methods to predict miRNA target genes with even more accuracy than what is currently possible. A number of groups have reported expression profile data for miRNAs in different tissues or developmental stages (Aravin *et al.*, 2003; Barad *et al.*, 2004; Baskerville and Bartel, 2005; Chen *et al.*, 2005). These data may be combined with expression profile data from targeted mRNAs to detect potential *in vivo* interactions between miRNAs and their target genes. High-throughput experimental strategies such as microarray analysis have been developed and applied to miRNA target analysis, raising expectations for large-scale analysis of miRNA target sites (Barad *et al.*, 2004; Baskerville and Bartel, 2005; Wang and Wang, 2006). Additionally, "knockdown" or "knockout" genetic tests of miRNAs or their target genes may be crucial for gaining a better understanding of the correlation between the function of miRNAs and specific phenomenon.

Once miRNA targets can be predicted with a fair degree of accuracy, the next step may be to work to understand the roles of miRNAs in living systems. To do this, gaining a comprehensive view of gene expression systems may be crucial. Computational approaches, which excel in the handling of genomic, transcriptomic, and proteomic data, should provide invaluable tools for identification of the relative position of miRNAs in various biological networks. Analysis of miRNAs and their target genes in the context of additional functional genomic data is expected to shed light on the potentially diverse and important biological functions of miRNAs within living systems.

ACKNOWLEDGMENTS

The authors acknowledge Nozomu Yachie, Koji Numata, and Dr. Rintaro Saito of Keio University for their helpful discussions.

REFERENCES

Ambros, V., Bartel, B., Bartel, D. P., Burge, C. B., Carrington, J. C., Chen, X., Dreyfuss, G., Eddy, S. R., Griffiths-Jones, S., Marshall, M., Matzke, M., Ruvkun, G., and Tuschl, T. (2003). A uniform system for microRNA annotation. *RNA* **9,** 277–279.

Aravin, A. A., Lagos-Quintana, M., Yalcin, A., Zavolan, M., Marks, D., Snyder, B., Gaasterland, T., Meyer, J., and Tuschl, T. (2003). The small RNA profile during *Drosophila melanogaster* development. *Dev. Cell* **5**, 337–350.

Barad, O., Meiri, E., Avniel, A., Aharonov, R., Barzilai, A., Bentwich, I., Einav, U., Gilad, S., Hurban, P., Karov, Y., Lobenhofer, E. K., Sharon, E., Shiboleth, Y. M., Shtutman, M., Bentwich, Z., and Einat, P. (2004). MicroRNA expression detected by oligonucleotide microarrays: System establishment and expression profiling in human tissues. *Genome Res.* **14**, 2486–2494.

Bartel, D. P. (2004). MicroRNAs: genomics, biogenesis, mechanism, and function. *Cell* **116**, 281–297.

Baskerville, S., and Bartel, D. P. (2005). Microarray profiling of microRNAs reveals frequent coexpression with neighboring miRNAs and host genes. *RNA* **11**, 241–247.

Bennasser, Y., Le, S. Y., Yeung, M. L., and Jeang, K. T. (2004). HIV-1 encoded candidate micro-RNAs and their cellular targets. *Retrovirology* **1**, 43.

Bentwich, I. (2005). Prediction and validation of microRNAs and their targets. *FEBS Lett.* **579**, 5904–5910.

Bonnet, E., Wuyts, J., Rouze, P., and Van de Peer, Y. (2004). Detection of 91 potential conserved plant microRNAs in *Arabidopsis thaliana* and *Oryza sativa* identifies important target genes. *Proc. Natl. Acad. Sci. USA* **101**, 11511–11516.

Bray, N., Dubchak, I., and Pachter, L. (2003). AVID: A global alignment program. *Genome Res.* **13**, 97–102.

Brennecke, J., Stark, A., Russell, R. B., and Cohen, S. M. (2005). Principles of microRNA-target recognition. *PLoS Biol.* **3**, e85.

Brown, J. R., and Sanseau, P. (2005). A computational view of microRNAs and their targets. *Drug Discov. Today* **10**, 595–601.

Burgler, C., and Macdonald, P. M. (2005). Prediction and verification of microRNA targets by MovingTargets, a highly adaptable prediction method. *BMC Genomics* **6**, 88.

Chan, C. S., Elemento, O., and Tavazoie, S. (2005). Revealing posttranscriptional regulatory elements through network-level conservation. *PLoS Comput. Biol.* **1**, e69.

Chen, P. Y., Manninga, H., Slanchev, K., Chien, M., Russo, J. J., Ju, J., Sheridan, R., John, B., Marks, D. S., Gaidatzis, D., Sander, C., Zavolan, M., and Tuschl, T. (2005). The developmental miRNA profiles of zebrafish as determined by small RNA cloning. *Genes Dev.* **19**, 1288–1293.

Couronne, O., Poliakov, A., Bray, N., Ishkhanov, T., Ryaboy, D., Rubin, E., Pachter, L., and Dubchak, I. (2003). Strategies and tools for whole-genome alignments. *Genome Res.* **13**, 73–80.

Didiano, D., and Hobert, O. (2006). Perfect seed pairing is not a generally reliable predictor for miRNA-target interactions. *Nat. Struct. Mol. Biol.* **13**, 849–851.

Doench, J. G., and Sharp, P. A. (2004). Specificity of microRNA target selection in translational repression. *Genes Dev.* **18**, 504–511.

Enright, A. J., John, B., Gaul, U., Tuschl, T., Sander, C., and Marks, D. S. (2003). MicroRNA targets in *Drosophila*. *Genome Biol.* **5**, R1.

Gerlach, W., and Giegerich, R. (2006). GUUGle: a utility for fast exact matching under RNA complementary rules including G-U base pairing. *Bioinformatics* **22**, 762–764.

Griffiths-Jones, S. (2004). The microRNA Registry. *Nucleic Acids Res.* **32**, D109–D111.

Griffiths-Jones, S., Grocock, R. J., van Dongen, S., Bateman, A., and Enright, A. J. (2006). miRBase: microRNA sequences, targets and gene nomenclature. *Nucleic Acids Res.* **34**, D140–D144.

Grun, D., Wang, Y. L., Langenberger, D., Gunsalus, K. C., and Rajewsky, N. (2005). MicroRNA target predictions across seven *Drosophila* species and comparison to mammalian targets. *PLoS Comput. Biol.* **1**, e13.

Gustafson, A. M., Allen, E., Givan, S., Smith, D., Carrington, J. C., and Kasschau, K. D. (2005). ASRP: The Arabidopsis Small RNA Project Database. *Nucleic Acids Res.* **33**, D637–D640.

Hariharan, M., Scaria, V., Pillai, B., and Brahmachari, S. K. (2005). Targets for human encoded microRNAs in HIV genes. *Biochem. Biophys. Res. Commun.* **337**, 1214–1218.

Hofacker, I. L. (2003). Vienna RNA secondary structure server. *Nucleic Acids Res.* **31**, 3429–3431.

Hsu, P. W., Huang, H. D., Hsu, S. D., Lin, L. Z., Tsou, A. P., Tseng, C. P., Stadler, P. F., Washietl, S., and Hofacker, I. L. (2006). miRNAMap: Genomic maps of microRNA genes and their target genes in mammalian genomes. *Nucleic Acids Res.* **34**, D135–D139.

John, B., Enright, A. J., Aravin, A., Tuschl, T., Sander, C., and Marks, D. S. (2004). Human microRNA targets. *PLoS Biol.* **2**, e363.

John, B., Sander, C., and Marks, D. S. (2006). Prediction of human microRNA targets. *Methods Mol. Biol.* **342**, 101–113.

Jones-Rhoades, M. W., and Bartel, D. P. (2004). Computational identification of plant microRNAs and their targets, including a stress-induced miRNA. *Mol. Cell* **14**, 787–799.

Karolchik, D., Baertsch, R., Diekhans, M., Furey, T. S., Hinrichs, A., Lu, Y. T., Roskin, K. M., Schwartz, M., Sugnet, C. W., Thomas, D. J., Weber, R. J., Haussler, D., *et al.* (2003). The UCSC Genome Browser Database. *Nucleic Acids Res.* **31**, 51–54.

Kim, S. K., Nam, J. W., Rhee, J. K., Lee, W. J., and Zhang, B. T. (2006). miTarget: microRNA target-gene prediction using a support vector machine. *BMC Bioinform.* **7**, 411.

Kiriakidou, M., Nelson, P. T., Kouranov, A., Fitziev, P., Bouyioukos, C., Mourelatos, Z., and Hatzigeorgiou, A. (2004). A combined computational-experimental approach predicts human microRNA targets. *Genes Dev.* **18**, 1165–1178.

Krek, A., Grun, D., Poy, M. N., Wolf, R., Rosenberg, L., Epstein, E. J., MacMenamin, P., da Piedade, I., Gunsalus, K. C., Stoffel, M., and Rajewsky, N. (2005). Combinatorial microRNA target predictions. *Nat. Genet.* **37**, 495–500.

Lall, S., Grun, D., Krek, A., Chen, K., Wang, Y. L., Dewey, C. N., Sood, P., Colombo, T., Bray, N., Macmenamin, P., Kao, H. L., Gunsalus, K. C., Pachter, L., Piano, F., and Rajewsky, N. (2006). A genome-wide map of conserved microRNA targets in *C. elegans*. *Curr. Biol.* **16**, 460–471.

Lecellier, C. H., Dunoyer, P., Arar, K., Lehmann-Che, J., Eyquem, S., Himber, C., Saib, A., and Voinnet, O. (2005). A cellular microRNA mediates antiviral defense in human cells. *Science* **308**, 557–560.

Lewis, B. P., Burge, C. B., and Bartel, D. P. (2005). Conserved seed pairing, often flanked by adenosines, indicates that thousands of human genes are microRNA targets. *Cell* **120**, 15–20.

Lewis, B. P., Shih, I. H., Jones-Rhoades, M. W., Bartel, D. P., and Burge, C. B. (2003). Prediction of mammalian microRNA targets. *Cell* **115**, 787–798.

Li, X., and Zhang, Y. Z. (2005). Computational detection of microRNAs targeting transcription factor genes in *Arabidopsis thaliana*. *Comput. Biol. Chem.* **29**, 360–367.

Lim, L. P., Lau, N. C., Garrett-Engele, P., Grimson, A., Schelter, J. M., Castle, J., Bartel, D. P., Linsley, P. S., and Johnson, J. M. (2005). Microarray analysis shows that some microRNAs downregulate large numbers of target mRNAs. *Nature* **433**, 769–773.

Lin, S. Y., Johnson, S. M., Abraham, M., Vella, M. C., Pasquinelli, A., Gamberi, C., Gottlieb, E., and Slack, F. J. (2003). The C elegans hunchback homolog, hbl-1, controls temporal patterning and is a probable microRNA target. *Dev. Cell* **4**, 639–650.

Llave, C., Xie, Z., Kasschau, K. D., and Carrington, J. C. (2002). Cleavage of Scarecrow-like mRNA targets directed by a class of *Arabidopsis* miRNA. *Science* **297**, 2053–2056.

Mathews, D. H., Sabina, J., Zuker, M., and Turner, D. H. (1999). Expanded sequence dependence of thermodynamic parameters improves prediction of RNA secondary structure. *J. Mol. Biol.* **288**, 911–940.

Miranda, K. C., Huynh, T., Tay, Y., Ang, Y. S., Tam, W. L., Thomson, A. M., Lim, B., and Rigoutsos, I. (2006). A pattern-based method for the identification of microRNA binding sites and their corresponding heteroduplexes. *Cell* **126**, 1203–1217.

Pfeffer, S., Zavolan, M., Grasser, F. A., Chien, M., Russo, J. J., Ju, J., John, B., Enright, A. J., Marks, D., Sander, C., and Tuschl, T. (2004). Identification of virus-encoded micro-RNAs. *Science* **304,** 734–736.

Rajewsky, N. (2006). microRNA target predictions in animals. *Nat. Genet.* **38**(Suppl.), S8–S13.

Rajewsky, N., and Socci, N. D. (2004). Computational identification of microRNA targets. *Dev. Biol.* **267,** 529–535.

Rehmsmeier, M. (2006). Prediction of microRNA targets. *Methods Mol. Biol.* **342,** 87–99.

Rehmsmeier, M., Steffen, P., Hochsmann, M., and Giegerich, R. (2004). Fast and effective prediction of microRNA/target duplexes. *RNA* **10,** 1507–1517.

Rhoades, M. W., Reinhart, B. J., Lim, L. P., Burge, C. B., Bartel, B., and Bartel, D. P. (2002). Prediction of plant microRNA targets. *Cell* **110,** 513–520.

Robins, H., Li, Y., and Padgett, R. W. (2005). Incorporating structure to predict micro-RNA targets. *Proc. Natl. Acad. Sci. USA* **102,** 4006–4009.

Rusinov, V., Baev, V., Minkov, I. N., and Tabler, M. (2005). MicroInspector: A web tool for detection of miRNA binding sites in an RNA sequence. *Nucleic Acids Res.* **33,** W696–W700.

Saetrom, O., Snove, O., Jr., and Saetrom, P. (2005). Weighted sequence motifs as an improved seeding step in microRNA target prediction algorithms. *RNA* **11,** 995–1003.

Sethupathy, P., Corda, B., and Hatzigeorgiou, A. G. (2006). TarBase: A comprehensive database of experimentally supported animal microRNA targets. *RNA* **12,** 192–197.

Shahi, P., Loukianiouk, S., Bohne-Lang, A., Kenzelmann, M., Kuffer, S., Maertens, S., Eils, R., Grone, H. J., Gretz, N., and Brors, B. (2006). Argonaute—a database for gene regulation by mammalian microRNAs. *Nucleic Acids Res.* **34,** D115–D118.

Stark, A., Brennecke, J., Bushati, N., Russell, R. B., and Cohen, S. M. (2005). Animal microRNAs confer robustness to gene expression and have a significant impact on 3′ UTR evolution. *Cell* **123,** 1133–1146.

Stark, A., Brennecke, J., Russell, R. B., and Cohen, S. M. (2003). Identification of *Drosophila* microRNA targets. *PLoS Biol.* **1,** E60.

Vella, M. C., Reinert, K., and Slack, F. J. (2004). Architecture of a validated microRNA:: target interaction. *Chem. Biol.* **11,** 1619–1623.

Wang, X., and Wang, X. (2006). Systematic identification of microRNA functions by combining target prediction and expression profiling. *Nucleic Acids Res.* **34,** 1646–1652.

Wang, X. J., Reyes, J. L., Chua, N. H., and Gaasterland, T. (2004). Prediction and identification of *Arabidopsis thaliana* microRNAs and their mRNA targets. *Genome Biol.* **5,** R65.

Washietl, S., Hofacker, I. L., and Stadler, P. F. (2005). Fast and reliable prediction of noncoding RNAs. *Proc. Natl. Acad. Sci. USA* **102,** 2454–2459.

Watanabe, Y., Yachie, N., Numata, K., Saito, R., Kanai, A., and Tomita, M. (2006). Computational analysis of microRNA targets in *Caenorhabditis elegans. Gene* **365,** 2–10.

Waterman, M. S., and Eggert, M. (1987). A new algorithm for best subsequence alignments with application to tRNA-rRNA comparisons. *J. Mol. Biol.* **197,** 723–728.

Wuchty, S., Fontana, W., Hofacker, I. L., and Schuster, P. (1999). Complete suboptimal folding of RNA and the stability of secondary structures. *Biopolymers* **49,** 145–165.

Yoon, S., and De Micheli, G. (2006). Computational identification of microRNAs and their targets. *Birth Defects Res. C Embryo Today* **78,** 118–128.

Zhang, Y. (2005). miRU: An automated plant miRNA target prediction server. *Nucleic Acids Res.* **33,** W701–W704.

MicroRNA EXPRESSION, MATURATION, AND FUNCTIONAL ANALYSIS

In Vitro and *In Vivo* Assays for the Activity of Drosha Complex

Yoontae Lee *and* V. Narry Kim

Contents

Abstract

MicroRNA (miRNA) genes are transcribed into long primary transcripts (pri-miRNAs) that get processed into mature miRNAs of about 22 nt in length by two different ribonuclease (RNase) III enzymes, Drosha and Dicer. Various experimental protocols have been developed and modified for genetic and biochemical analyses for microRNA processing. Here we describe the methods for the analysis of pri-miRNA processing that is mediated by Drosha and its cofactor, DiGeorge Syndrome Critical Region Gene 8 (DGCR8).

1. Introduction

miRNAs are single-stranded RNAs of 19 to 25 nt in length generated from endogenous hairpin-shaped transcripts (Ambros *et al.*, 2003; Bartel, 2004; Kloosterman and Plasterk, 2006). Accumulating evidence indicates that miRNAs function in diverse regulatory pathways, such as developmental timing control, organ development, cell differentiation, apoptosis, cell proliferation, and tumorigenesis. To unravel the regulatory networks operated by miRNAs, it is necessary to elucidate how the expression of miRNA itself is regulated in cells.

School of Biological Sciences, Seoul National University, Seoul, Korea

Methods in Enzymology, Volume 427
ISSN 0076-6879, DOI: 10.1016/S0076-6879(07)27005-3

Genesis of most miRNAs begins with transcription by RNA polymerase II (pol II) (Cai *et al.*, 2004; Kim, 2005; Lee *et al.*, 2004). Some miRNAs associated with Alu repeats can be transcribed by RNA polymerase III (Borchert *et al.*, 2006) (Fig. 5.1). Like other pol II transcripts, pri-miRNAs are both capped and polyadenylated, although these end structures are not required for miRNA processing. Pri-miRNAs are usually over several kilobases long and often exceed 10 kilobases. Thus, precise processing is critical to ensure the production of functional mature miRNA of ~22 nt. Processing is initiated in the nucleus in a process known as cropping, and this is catalyzed by a member of the ribonuclease III family (RNase III) named Drosha (Lee et al, 2003). Drosha requires a cofactor that is known as

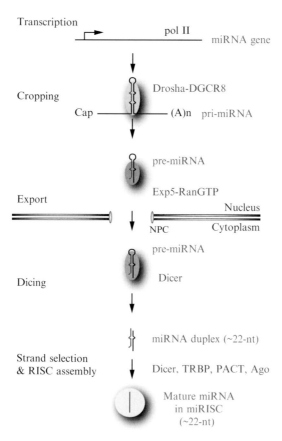

Figure 5.1   Current model for microRNA maturation. (See color insert.)

DGCR8 or Pasha for its activity (Denli *et al.*, 2004; Gregory *et al.*, 2004; Han *et al.*, 2004; Landthaler *et al.*, 2004). The Drosha–DGCR8 complex cleaves pri-miRNA near the base of the hairpin structure and releases a ∼65-nt-small hairpin called pre-miRNA. Pre-miRNAs then exit the nucleus via the action of exportin-5 (Exp5)(Bohnsack *et al.*, 2004; Lund *et al.*, 2004; Yi *et al.*, 2003). Upon export, pre-miRNAs are further cleaved at close to the terminal loop, yielding the ∼22 nt miRNA duplex. This reaction is carried out by Dicer, a cytoplasmic RNase III protein (Bernstein *et al.*, 2001; Grishok *et al.*, 2001; Hutvagner *et al.*, 2001; Ketting *et al.*, 2001; Knight and Bass, 2001). One strand of the duplex is subsequently loaded onto the RNA-induced silencing complex (RISC) that executes the RNA silencing procedure (Khvorova *et al.*, 2003; Schwarz *et al.*, 2003).

Drosha is conserved only in animals (Filippov *et al.*, 2000; Fortin *et al.*, 2002; Wu *et al.*, 2000). Although plants express diverse miRNAs, they do not possess a Drosha homologue, suggesting that miRNA biogenesis pathway is distinct in plants. In *Arabidopsis*, there are four Dicer homologues, among which Dicer-like 1 (DCL1) is known to be responsible for miRNA processing in the nucleus (Kurihara and Watanabe, 2004). It remains unclear how DCL1 recognizes its substrates and determines the cleavage sites.

Drosha was originally described as a gene located next to the RNase H gene in *Drosophila* genome (Filippov *et al.*, 2000). An early study on human Drosha homologue (initially referred to as human RNase III protein) showed Drosha as a nuclear protein involved in pre-rRNA processing because the knockdown of Drosha by antisense oligonucleotides resulted in the moderate accumulation of rRNA precursors (Wu *et al.*, 2000). It remains to be determined, however, whether Drosha is directly involved in pre-rRNA processing. In fact, we could not detect the accumulation of rRNA precursors when Drosha was depleted by RNAi (Han and Kim, unpublished data). Thus far, no other endogenous RNA species apart from pri-miRNAs has been found to be processed by Drosha.

Drosha is a large nuclear protein of 130 to 160 kDa (160 kDa in humans) with multiple domains (Fig. 5.2A). Drosha contains two tandem RNase III domains (RIIIDs) and a double-stranded RNA binding domain (dsRBD). The N-terminal portion of Drosha contains a proline-rich region as well as a serine/arginine-rich region. Deletions of the N-terminal region that remove the P-rich region (ΔN220) and most of the RS-rich region (ΔN390) did not affect the enzymatic activity (Han *et al.*, 2004). The RS-rich region appears to possess the nuclear localization signal because the mutant ΔN390 fails to localize to the nucleus (Han and Kim, unpublished result). The remainder of the mutants (ΔN490, ΔC432, and ΔC114) turned out to be inactive in our assay, indicating that the middle region as well as the RIIIDs and the dsRBD are required for pri-miRNA processing (Han *et al.*, 2004).

A   B

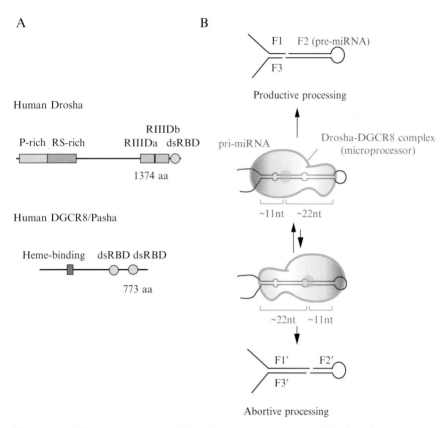

Figure 5.2 The microprocessor. (A) Schematic presentation of the domain structures of human Drosha and DGCR8. (B) The model for substrate recognition and cleavage site selection by the microprocessor. (See color insert.)

Drosha forms a large complex that weighs ~500 kDa in *Drosophila* (Denli *et al.*, 2004) or ~650 kDa in humans (Gregory *et al.*, 2004; Han *et al.*, 2004). In this complex known as the microprocessor (Denli *et al.*, 2004; Gregory *et al.*, 2004), Drosha interacts with its cofactor, DGCR8/Pasha (Denli *et al.*, 2004; Gregory *et al.*, 2004; Han *et al.*, 2004; Landthaler *et al.*, 2004) (see Fig. 5.2A). Neither recombinant DGCR8/Pasha nor Drosha alone is active in pri-miRNA processing, whereas combining these two proteins restores pri-miRNA processing activity, indicating that both proteins are essential for cropping reaction (Gregory *et al.*, 2004; Han *et al.*, 2004).

The human DGCR8 gene is located on chromosome 22q11 and is expressed ubiquitously from fetus to adult (Shiohama *et al.*, 2003). Monoallelic deletion of this genomic region is associated with DiGeorge syndrome (Goldberg *et al.*, 1993) although it remains unknown whether DGCR8 plays any role in DiGeorge syndrome. It was shown that DGCR8 knockout is lethal at an early embryonic stage (Wang *et al.*, 2007). The analysis using knockout embryonic stem (ES) cells showed that knockout ES cells are depleted of almost all miRNAs, confirming that DGCR8 is an essential component in miRNA processing. Interestingly, DGCR8-null ES cells are defective in cell cycle progression and cell differentiation.

DGCR8/Pasha is a $\sim$120 kDa protein that contains two dsRBDs at the C-terminus and the heme-binding domain in its middle region (see Fig. 5.2A). A relatively small region of the human DGCR8 protein (residues 484 through 750) containing the dsRBDs and the C-terminal Drosha-binding domain is sufficient to support pri-miRNA processing *in vitro* (Han *et al.*, 2006). The domain responsible for Drosha binding is located at the C-terminus of DGCR8 (Han *et al.*, 2006). DGCR8 binds to the middle region and the RIIIDs of the Drosha protein (Han *et al.*, 2004). This interaction appears to stabilize the Drosha protein (Han *et al.*, 2006). The N-terminal region is not critical for processing, but is important for nuclear localization of DGCR8 (Han *et al.*, 2006). It was shown that heme is bound to DGCR8 through Cys352 and activates DGCR8 possibly through promoting oligomerization (Faller *et al.*, 2007).

DGCR8/Pasha is capable of direct interaction with pri-miRNA (Han *et al.*, 2006). Drosha appears to have much lower affinity to the substrates. Thus, DGCR8 may be the primary factor that recognizes the pri-miRNA structure.

A typical metazoan pri-miRNA consists of a stem of $\sim$33 bp, with a terminal loop and flanking ssRNA segments. Drosha cleaves both natural and artificial substrates at a site $\sim$11 bp away from the ssRNA–dsRNA junction (see Fig. 5.2B). Manipulating the distance from the ssRNA–dsRNA junction affects cleavage site selection. A series of mutagenesis and biochemical analyses revealed that the flanking ssRNA segments are critical for processing to occur (Han *et al.*, 2006). The terminal loop is less significant to pri-miRNA processing *in vitro*. In fact, the flanking ssRNA segments are important for the binding to DGCR8. The dsRNA stem of over 33 bp is also required for efficient binding. Thus, DGCR8 may function as the molecular anchor that measures the distance of the dsRNA–ssRNA junction.

Because some large terminal loops can be seen as unstructured ssRNA segments, microprocessor can recognize the terminal loop as ssRNA and binds to the stem-loop in an opposite orientation (see Fig. 5.2B). In this case, cleavage can occur at an alternative site (at $\sim$11 bp from the terminal loop) (see Fig. 5.2B). The processing at this alternative site would be "abortive" because the cleavage product does not contain miRNA sequences in full. The efficiency of abortive processing is usually much lower than that of productive processing in pri-miRNA processing (see Fig. 5.2B).

2. ASSAY METHODS

2.1. *In vitro* analysis of pri-miRNA processing

2.1.1. Preparation of radiolabeled pri-miRNA transcript

The RNA substrate for a cropping reaction can be prepared by *in vitro* transcription. The DNA segment containing pri-miRNA sequences can be amplified from genomic DNA by PCR. The PCR product is then inserted into a plasmid containing the T7 promoter (or SP6 promoter). The resulting plasmid needs to be linearized at the downstream of the pri-miRNA sequences prior to transcription. Alternatively, the PCR product can be used directly for *in vitro* transcription. In this case, the forward PCR primer should contain the promoter sequences (usually the T7 promoter) to drive transcription directly from the PCR product.

When designing the PCR primers, one has to decide which region around the miRNA should be included to achieve efficient processing. In general, all the *cis*-acting element(s) for processing are located close to the stem-loop (Chen *et al.*, 2004; Han *et al.*, 2006; Lee *et al.*, 2003; Zeng and Cullen, 2005; Zeng *et al.*, 2005). By the rule of thumb, we amplify a genomic segment containing the miRNA stem-loop plus the surrounding sequences ~100 bp from each side of the stem-loop. This should normally include all the *cis*-acting element(s) recognized by the Drosha–DGCR8 complex.

Materials

1. Template plasmid (1 μg/μl) or PCR product (30 ng/μl); The plasmid must be linearized at a restriction site located downstream of the pri-miRNA sequences. The PCR products are subcloned into a cloning vector containing the T7 or SP6 promoter (e.g., pGEM-T-easy, Promega). It is also possible to use the PCR product directly as the template in transcription reaction if the T7 promoter sequence is included in the forward PCR primer. For efficient transcription, the promoter sequences should be followed by two consecutive Gs.
2. NTP mixture (10 mM ATP, 10 mM GTP, 10 mM CTP, 1 mM UTP)
3. α-^{32}P-UTP (20 μCi/μl, 800 mCi/mmol)
4. T7 or SP6 RNA polymerase
5. 10$\times$ transcription buffer (use the buffer provided with the polymerase)
6. RNase inhibiter (RNasin or equivalent, 40 U/μl)
7. RNase-free water
8. Hoeffer gel apparatus (SE600, 18 $\times$ 16 cm) or equivalent, plates, combs (0.75 mm, 10 wells), spacers (0.75 mm), and a power supply
9. TBE stock solution (5$\times$): dissolve 54 g of Tris base, 27.5 g of boric acid, and 4 ml of 0.5 M EDTA, pH 8.0 in distilled water to make 1 l.

10. 6% urea–polyacrylamide stock solution: mix 57 g acrylamide, 3 g Bis-acrylamide, 100 ml of 5× TBE stock solution (final 0.5×) and 420.42 g urea (final 7 M) in distilled water to make 1 l. When preparing urea-polyacrylamide stock solution, dissolve urea in water at 60° because this reaction is an endothermic reaction.
11. 20% ammonium persulfate (APS) solution
12. N,N,N',N'-tetramethylethylenediamine (TEMED)
13. Phenol:chloroform:isoamyl alcohol (IAA) (25:24:1), pH 4.5
14. 3 M sodium acetate, pH 5.5
15. Glycogen (5 mg/ml)
16. TE buffer: 10 mM Tris-HCl, pH 7.5 and 1 mM EDTA
17. RNA elution buffer: 0.3 M sodium acetate, pH 5.5 and 2% SDS
18. RNA loading buffer: 95% deionized formamide, 0.025% bromophenol blue, 0.025% xylene cyanol, 5 mM EDTA, and 0.025% SDS
19. Ethanol: 100 and 75%
20. Microcentrifuge with cooling
21. Thermoblocks at 37°/95°
22. Autoradiography films

Procedure

1. Mix the following at room temperature (rt): 1 μl of template DNA, 2 μl of 10× transcription buffer, 2 μl of NTP mixture (10 mM ATP, 10 mM GTP, 10 mM CTP, 1 mM UTP), 0.5 μl of RNase inhibiter (40 u/μl), 1.5 μl of α-^{32}P-UTP (10 μCi/μl, 800 Ci/mmol), 1 μl of RNA polymerase (T7 or SP6), and 12 μl of RNase-free water. Adjust the total volume to 20 μl.
2. Incubate at 37° for 3 h. In the meantime, it is recommended that the gel is prepared (steps 11–13).
3. Add 220 μl of TE buffer to the preceding reaction mixture.
4. Add 240 μl of phenol:chloroform:IAA (25:24:1) and vortex for 30 sec. Centrifuge for 5 min at rt and take the upper layer.
5. Add 160 μl of 5-M ammonium acetate, 1 μl of glycogen, and 1 ml of 100% ethanol. Mix and leave the tube at −80° for at least 20 min (or 5 min in dry ice-methanol mix or overnight at −20°).
6. Centrifuge at full speed (at least 12,000 rpm) at 4° for 15 min.
7. Remove the supernatant carefully so as not to disturb the RNA pellet.
8. Wash the pellet with 500 μl of 75% ethanol, centrifuge at 12,000 rpm at 4° for 5 min, and decant the ethanol. Quick-spin the tube and carefully remove the residual alcohol.
9. Air-dry the pellet for ∼5 min (be careful not to overdry).
10. Resuspend the pellet in 20 μl of RNA loading buffer. Boil this RNA sample at 95° for 5 min. Quick-spin the tube and leave it at rt until loading.
11. Assemble a gel cast.

12. Mix 20 ml of 6% urea-polyacrylamide stock solution, 100 μl of 20% APS, and 20 μl of TEMED. Pour this mixture into the gel cast immediately and insert a comb as fast as possible because the gel solidifies in a few minutes.
13. Pre-run at 350 V for at least 60 min. Running buffer is $0.5 \times$ TBE.
14. Load the RNA sample (step 10) on 6% urea-polyacrylamide gel and run at 350 V until bromophenol blue reaches the bottom of the gel.
15. Dissemble the gel cast and remove one of the glass plates from the gel. Place an X-ray film on the gel for 30 to 60 sec. Make sure to mark the position and orientation of the gel on the film so that the gel can be aligned with the film once the film is developed. The radiolabeled transcript will appear as a strong band on the developed film.
16. Align the film with the gel and cut out the gel slice containing the labeled transcript. Put the gel slice in 350 μl of RNA elution buffer.
17. Incubate overnight at 42°.
18. Transfer the supernatant (about 300 μl) to a fresh tube.
19. Add 100 μl of RNA elution buffer to the gel slice and vortex. Remove the supernatant and add it to the previous supernatant.
20. To the supernatant, add 1 μl of glycogen and 1 ml of 100% ethanol.
21. Mix and place the tube at $-80°$ for at least 20 min (or 5 min in dry ice-methanol mix, or overnight at $-20°$).
22. Spin at full speed at 4° for 15 min.
23. Wash the pellet with 500 μl of 75% ethanol, centrifuge at 12,000 rpm at 4° for 5 min, and decant the ethanol. Quick-spin the tube and carefully remove the residual alcohol.
24. Air-dry the pellet. Be careful not to overdry the pellet.
25. Count counts per minute (cpm) and resuspend in RNase-free TE buffer.

2.1.2. Preparation of the microprocessor
Materials

1. Mammalian expression plasmids containing human Drosha cDNA fused with FLAG epitope at its C-terminus (pCK-Drosha-FLAG) or DGCR8 fused with FLAG epitope at its N-terminus (pCK-FLAG-DGCR8)
2. HEK293T cells, culture medium (DMEM supplemented with 10% FBS), culture equipments, transfection reagent (calcium phosphate method is the most economic and efficient choice for this experiment)
3. Phosphate-buffered saline (PBS): dissolve 8 g NaCl, 0.2 g KCl, 1.44 g $Na_2HPO_4 \cdot 2H_2O$, and 0.2 g KH_2PO_4 in water. Check pH (should be 7.2), set volume to 1 l, and autoclave
4. Lysis buffer: 20 mM Tris-HCl (pH 8.0), 100 mM KCl, and 0.2 mM EDTA
5. Anti-FLAG antibody conjugated agarose (Anti-FLAG M2 Affinity Gel Freezer-Safe, Sigma)
6. Sonicator (Sonics, VC130)

Procedure
Preparation of HEK293T whole cell extract

1. Grow HEK293T cells on 100-mm dish and transfect the cells with 8 μg of pCK-Drosha-Flag. To obtain stronger processing activity, it is recommended that 2 μg of DGCR8 expression plasmid is cotransfected along with Drosha expression plasmid.
2. Two days later, remove the media and rinse the cells with 5 ml of ice-cold PBS.
3. Add 1 ml of ice-cold PBS, collect the cells by pipetting, and transfer to an Eppendorf tube.
4. Centrifuge at 6000 rpm at 4° for 5 min.
5. Decant the PBS and resuspend the cell pellet in 500 μl of lysis buffer.
6. Sonicate for 1 min with 5 sec pulses at 30% amplitude.
7. Centrifuge at 12,000 rpm at 4° for 15 min.
8. Transfer the supernatant to a fresh tube. This is the HEK293T whole cell extract.

Immunoprecipitation of the microprocessor

1. Prepare HEK293T whole cell extract as just described using HEK293T cells transfected with the Drosha–FLAG or FLAG-DGCR8 expression plasmid.
2. Take ~10 μl (bead volume) of anti-FLAG antibody conjugated beads and wash the beads by adding 1 ml of lysis buffer, mixing, centrifugation, and decanting the supernatant. Repeat washing once.
3. Add 500 μl of HEK293T whole cell extract to the washed beads.
4. Rotate the tube at 4° for 1 h.
5. Centrifuge at 8000 rpm at 4° for 30 sec.
6. Remove the supernatant, add 1 ml of lysis buffer, and wash the beads by inverting the tube 6 to 7 times.
7. Centrifuge at 8000 rpm at 4° for 30 sec.
8. Repeat steps 6 and 7 four times. Change the tube for the last washing.
9. Carefully remove the residual buffer so that the residual volume becomes approximately 15 μl. Keep the tube on ice until this is used for assay. (We do not freeze or store the immunoprecipitate on ice for more than 3 h. Basically, we prepare the immunoprecipitate freshly for each assay.)

2.1.3. *In vitro* pri-miRNA processing assay
Materials

1. Radiolabeled pri-miRNA transcript
2. HEK293T whole cell extract or the agarose beads holding the immunoprecipitated microprocessor
3. 64 mM MgCl$_2$ (RNase III family proteins require magnesium ion for catalysis)
4. RNase inhibitor (RNasin or equivalent, 40 U/μl)

5. Hoeffer gel apparatus (18 × 16 cm) or equivalent, combs (0.75 mm, 15 wells), spacers (0.75 mm), and a power supply
6. Urea-polyacrylamide stock solution (12.5%): Dissolve 118.75 g acrylamide, 6.25 g Bis-acrylamide, 100 ml 5× TBE (final 0.5×) and 420.42 g Urea (final 7 M) in water to make 1 l.
7. APS solution: 20% dissolved in water
8. TEMED
9. RNA elution buffer: 0.3 M sodium acetate, pH 5.5 and 2% SDS
10. Phenol:chloroform:isoamyl alcohol (IAA) (25:24:1), pH 4.5
11. 3-M sodium acetate solution, pH 5.5
12. RNA loading buffer: 95% deionized formamide, 0.025% bromophenol blue, 0.025% xylene cyanol, 5 mM EDTA, and 0.025% SDS
13. Thermoblock at 37°/95°
14. Autoradiography films
15. Autoradiography cassettes with intensifying screens

Procedure

1. Mix the following: 3 μl of 64 mM MgCl$_2$ solution (final 6.4 mM), 3 μl of radiolabeled pri-miRNA ($10^4 \sim 10^5$ cpm), 0.75 μl of RNase inhibitor (final 1U/μl), 8.25 μl of RNase-free water, and 15 μl of HEK293T whole extract or immunoprecipitate (immobilized on the beads).
2. Incubate at 37° for 90 min.
3. Add 170 μl of RNA elution buffer to the reaction mixture.
4. Add 200 μl of phenol:chloroform:isoamylalcohol (25:24:1) and vortex for 30 sec.
5. Centrifuge for 5 min at rt and take the upper layer.
6. Add 20 μl of 3 M sodium acetate, 1 μl of glycogen, and 800 μl of 100% ethanol. Mix and leave at −80° for at least 20 min (or 5 min in dry ice-methanol mix, or overnight at −20°).
7. Centrifuge at maximum speed at 4° for 15 min.
8. Remove the supernatant carefully so as not to disturb the RNA pellet. Wash the pellet with 500 μl of 75% ethanol.
9. Air-dry the pellet.
10. Resuspend the pellet in 15 μl of RNA loading buffer.
11. Prepare 12.5% urea-polyacrylamide gel as previously described (see 2.1.1, Preparation of radiolabeled pri-miRNA transcript, *procedure,* steps 11–13).
12. Heat the RNA sample at 95° for 5 min and quick-spin the tube. Load 7.5 μl of RNA samples on the gel and run at 350 V until bromophenol blue reaches the bottom of the gel.
13. Wrap the gel in plastic wrap. Expose the gel on an X-ray film overnight at −80° and develop this film. The product of processing will appear as a band of 60 to 75 nt depending on the substrate used in the assay. For instance, pre-let-7a is 72 nt in length (Fig. 5.3B). Additional bands may appear, which usually correspond to the flanking sequences around

A   B

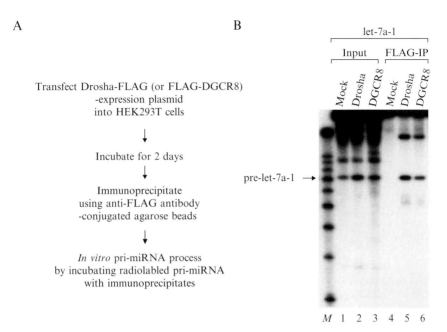

Figure 5.3 *In vitro* pri-miRNA processing assay. (A) Experimental scheme. (B) Typical results from processing assay. Pri-let-7a-1 was incubated with HEK293T whole cell extract (input) or immunoprecipitates (FLAG-IP). HEK293T cells had been transfected with pCK (Mock), pCK-Drosha-FLAG (Drosha), or pCK-FLAG-DGCR8 (DGCR8). Pre-let-7a-1 is produced when pri-miRNA is incubated with Drosha or DGCR8 immunoprecipitate or whole cell extract.

the cleavage sites. Contamination by other nuclease(s), nonspecific chemical cleavage reaction, or abortive processing can also result in unexpected cleavage products.

2.2. *In vivo* analysis of pri-miRNA processing

The function of the microprocessor can be examined *in vivo* by depleting Drosha or DGCR8 by RNAi. The effect of depletion on miRNA biogenesis can be examined by RT-PCR and Northern blot analysis. RT-PCR is the easiest choice to determine pri-miRNA level. The PCR primers are usually designed to bind ~100-nt away from the stem-loop so that the PCR product is 200 to 300 bp. Northern blot analysis is to detect mature miRNA as well as pre-miRNA. Both mature and pre-miRNA diminish after depletion of Drosha or DGCR8. However, it may take more than 6 days to observe a dramatic effect on mature miRNA because most mature miRNAs have long half-lives. For prolonged incubation, it is desirable to transfect siRNA twice by splitting the cells on the third day and repeating transfection on the fourth day.

2.2.1. Depletion of Drosha or DGCR8

Materials

1. Drosha-targeting siRNA duplex (siDrosha) used in our lab (20-μM stock);
SiRNA passenger strand: 5′-CGAGUAGGCUUCGUGACUUdTdT-3′
SiRNA guide strand: 5′-AAGUCACGAAGCCUACUCGdTdT-3′
2. DGCR8-targeting siRNA duplex (siDGCR8) used in our lab (20-μM stock);
SiRNA passenger strand: 5′-CAUCGGACAAGAGUGUGAUdTdT-3′
SiRNA guide strand: 5′-AUCACACUCUUGUCCGAUGdTdT-3′
3. Lipofectamine 2000 (Invitrogen)
4. FBS-free DMEM
5. DMEM supplemented with 10% FBS
6. DMEM supplemented with 20% FBS
7. Opti-MEM (Invitrogen)
8. HeLa cells

Procedure

1. One day before transfection, split HeLa cells in six-well plates and incubate in 2 ml of medium so that the cells become 70 to 80% confluent on the day of transfection. DMEM tissue culture medium is supplemented with 10% FBS.
2. Mix 4 μl of 20 μM siRNA duplex with 250 μl of Opti-MEM.
3. Mix 6 μl of lipofectamine 2000 with 250 μl of Opti-MEM and incubate at rt for 5 min.
4. Mix the siRNA solution (step 2) and the lipofectamin 2000 solution (step 3) and incubate at rt for 15 min.
5. While incubating the siRNA mixture, replace the media with 2 ml of FBS-free DMEM.
6. Add the final solution to HeLa cells.
7. After 6 h, add 2 ml of DMEM tissue culture medium supplemented with 20% FBS to the media.
8. After 3 days, extract total RNA or protein from HeLa cells and analyze the level of Drosha or DGCR8 by RT-PCR or Western blotting.

2.2.2. RT-PCR for pri-miRNA

Materials

1. 5 μg of total RNA extracted from HeLa cells treated with siRNA
2. Reverse transcriptase (Superscript II, Invitrogen)
3. Oligo-dT primer (Invitrogen)
4. RNase inhibitor (RNasin or equivalent, 40 U/μl)
5. Taq polymerase (rTaq, TAKARA)
6. dNTP mix (2.5 mM each)

7. PCR primers (10 μM)
8. PCR machine (T1 thermocycler, Biometra)

Procedure

1. Synthesize cDNA from HeLa total RNA using Superscript II (Invitrogen) and oligo-dT primer (Invitrogen) according to the manufacturer's protocol.
2. Carry out PCR using 2 μl of cDNA and 1 μl of PCR primers. A typical PCR condition is as follows:

Step 1: Denaturation at 95° for 3 min
Step 2: Denaturation at 95° for 30 sec followed by primer annealing for 30 sec at a suitable temperature depending on Tm value of the PCR primers and by elongation at 72° for 30 sec, 25 to 35 cycles
Step 3: 72° for 7 min

3. Analyze the PCR products on 2% agarose gel. The bands of the PCR product will intensify when Drosha or DGCR8 is depleted (Fig. 5.4B).

2.2.3. Northern blot analysis for miRNA

Materials

1. 50 μg of total RNA extracted from HeLa cells treated with siRNA
2. RNA loading buffer
3. Probe: DNA oligonucleotide complementary to the microRNA of interest (6 μM)
4. γ-^{32}P-ATP (10 μCi/μl, 6000 Ci/mmol)
5. T4-polynucleotide kinase (T4 PNK, TAKARA)
6. RNA size marker (Decade RNA marker, Ambion)
7. Positive charged nylon membrane (Zeta-probe GT membrane, Bio-Rad)
8. 3-M paper
9. Transfer unit: Hoeffer TE70 SemiPhor Semi-Dry Transfer Unit
10. 0.5× TBE
11. Hoeffer gel apparatus (18 × 16 cm) or equivalent, combs (1 mm, 15 wells), spacers (1 mm), and a power supply
12. Urea-polyacrylamide stock solution (12.5%): Dissolve 118.75 g acrylamide, 6.25 g Bis-acrylamide, 100 ml 5× TBE (final 0.5×), and 420.42 g Urea (final 7 M) in water to make 1 l.
13. APS solution: 20% dissolved in water
14. 3 M sodium acetate, pH 5.5
15. Glycogen (5 mg/ml)
16. TEMED
17. Ultraviolet (UV)-crosslinker
18. Hybridization chamber
19. Hybridization bottle
20. Hybridization buffer (Expresshyb solution, Clontech)
21. Salmon sperm DNA (10 mg/ml)

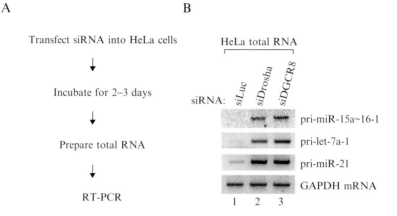

Figure 5.4 RT-PCR of pri-miRNA. (A) Experimental scheme. (B) Typical results. The siRNA duplex against luciferase (siLuc), Drosha (siDrosha), or DGCR8 (siDGCR8) was transfected into HeLa cells. After 72 h, total RNA was prepared and used for RT-PCR.

22. 20× SSC: Dissolve 175.3 g NaCl (final 3 M) and 88.2 g sodium citrate (final 0.3 M) in water to make 1 l. Adjust the pH to 7.0 with a solution of HCl and then sterilize by autoclaving.
23. 10% SDS solution
24. Washing solution I: 2× SSC, 0.05% SDS
25. Washing solution II: 0.1× SSC, 0.1% SDS
26. Glass tray
27. Autoradiography films
28. Autoradiography cassettes with intensifying screens

Procedure
Gel electrophoresis and blotting

1. Assemble a gel cast.
2. Mix 30 ml of 12.5% urea–polyacrylamide stock solution, 100 μl of 20% APS, and 20 μl of TEMED. Pour this mixture into the gel cast immediately and insert a comb as fast as possible because the gel solidifies in a few minutes.
3. Pre-run at 350 V for at least 60 min using 0.5× TBE as the running buffer.
4. While pre-running the gel, prepare RNA samples. Add 10 μl of RNA loading buffer to 50 μg of HeLa total RNA dissolved in 10 μl of TE buffer, and then boil this at 95° for 5 min.
5. Load the RNA sample (from step 4) on 12.5% pre-run urea-polyacrylamide gel and run at 350 V until bromophenol blue reaches the bottom of the gel. For RNA size marker, we use Decade RNA

marker from Ambion. The markers are end-labeled with γ-^{32}P-ATP and polynucleotide kinase according to the manufacturer's manual. The marker is loaded on a lane that is two lanes away from the samples. The amount of radioactivity of the marker should be 30 ~ 50 c.p.s.

6. Dissemble the gel cast and remove one of the glass plates from the gel. Place OHP film on the gel surface and detach the gel from the glass plate.
7. Transfer the gel to a glass tray filled with 200 ml of 0.5× TBE buffer.
8. Add a few drops of EtBr to the 0.5× TBE and stain the gel for 15 min on a rocker. Destain it with 200 ml of 0.5× TBE for 5 min. Examine the gel under UV for quality and quantity of the RNA.
9. Soak Zeta-Probe GT membrane in 0.5× TBE for 10 min. Soak four pieces of filter paper (3-M papers) in 0.5× TBE.
10. Assemble the transfer unit in the following order: two pieces of 3-M paper, Zeta-Probe GT membrane, gel, and two pieces of 3-M paper on the top.
11. Transfer at 250 mA at rt for 1 h.
12. Crosslink the transferred RNA to the membrane by UV-irradiation for 1 min 30 sec.
13. Sandwich the membrane between two dry 3-M papers and bake at 80° for 30 min.
14. Store at rt between filter papers until use.

Preparation of the probe

1. Mix the following: 3 μl of 6 μM DNA oligonucleotide probe, 2 μl of 10× T4 PNK buffer, 5 μl of γ-^{32}P-ATP, 1 μl of T4 PNK, and 9 μl of distilled water.
2. Incubate at 37° for 1 h.
3. Incubate at 68° for 10 min to inactivate T4 PNK. Add 180 μl of TE buffer, 20 μl of 3 M sodium acetate (pH 5.5), 1 μl of glycogen, and 800 μl of 100% EtOH. Incubate at −80° for at least 20 min.
4. Spin at 13,200 rpm for 15 min at 4°.
5. Wash the pellet by adding 500 μl of 75% EtOH and spin again for 5 min.
6. Carefully remove the liquid, air-dry the pellet, and store the pellet at −20°.
7. Immediately before use, dissolve the probe in 50 μl of TE. We usually get ~1 × 10^{7}cpm in total. If this is used in 5-ml hybridization buffer, the final radioactivity is ~0.2 × 10^7 cpm/ml.

Hybridization

1. Soak the membrane with 2× SSC.
2. Put the membrane into the hybridization bottle.
3. Put 5 ml of hybridization solution (pre-warmed to 37°) into the bottle.
4. Add 50 μl of salmon sperm DNA (boiled at 95° for 3 min and quick-chilled on ice for 1 min) and mix.

5. Incubate at 37° for 30 min for pre-hybridization with rotation in a hybridization chamber.
6. Replace the solution with 5 ml of fresh hybridization solution pre-warmed at 37°.
7. Boil and quick-chill 50 μl of the radiolabeled probe and 50 μl of salmon sperm DNA and add them to the hybridization solution and mix well.
8. Incubate at 37° for 1 h for hybridization with rotation in the hybridization chamber.
9. Rinse the membrane in the bottle with 5~10 ml of washing solution I and repeat once.
10. Transfer the membrane to a glass tray filled with 200 ml of washing solution I.
11. Shake the glass tray at rt for 30 min.
12. Repeat washing with washing solution I.
13. Wash with washing solution II for 15 min, twice.
14. Remove the residual liquid quickly and wrap the membrane in plastic wrap and expose to an X-ray film at −80°. The bands corresponding to pre-miRNA and mature miRNA get weaker when Drosha or DGCR8 is successfully depleted (Fig. 5.5).

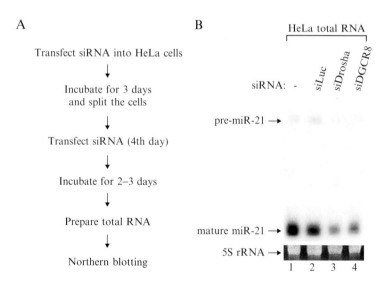

Figure 5.5 Northern blot analysis of pre-miRNA and mature miRNA. (A) Experimental scheme. (B) Typical results. RNAi was carried out by transfection of siRNA duplexes into HeLa cells. Both pre-miR-21 and mature miR-21 are detected by radiolabeled DNA oligo probe. 5S rRNA bands stained with ethidium bromide are presented as a loading control.

ACKNOWLEDGMENTS

We thank Jinju Han for proofreading the manuscript.

REFERENCES

Ambros, V., *et al.* (2003). A uniform system for microRNA annotation. *RNA* **9**, 277–279.

Bartel, D. P. (2004). MicroRNAs: Genomics, biogenesis, mechanism, and function. *Cell* **116**, 281–297.

Bernstein, E., Caudy, A. A., Hammond, S. M., and Hannon, G. J. (2001). Role for a bidentate ribonuclease in the initiation step of RNA interference. *Nature* **409**, 363–366.

Bohnsack, M. T., Czaplinski, K., and Gorlich, D. (2004). Exportin 5 is a RanGTP-dependent dsRNA-binding protein that mediates nuclear export of pre-miRNAs. *RNA* **10**, 185–191.

Borchert, G. M., Lanier, W., and Davidson, B. L. (2006). RNA polymerase III transcribes human microRNAs. *Nat. Struct. Mol. Biol.* **13**, 1097–1101.

Cai, X., Hagedorn, C. H., and Cullen, B. R. (2004). Human microRNAs are processed from capped, polyadenylated transcripts that can also function as mRNAs. *RNA* **10**, 1957–1966.

Chen, C. Z., Li, L., Lodish, H. F., and Bartel, D. P. (2004). MicroRNAs modulate hematopoietic lineage differentiation. *Science* **303**, 83–86.

Denli, A. M., Tops, B. B., Plasterk, R. H., Ketting, R. F., and Hannon, G. J. (2004). Processing of primary microRNAs by the Microprocessor complex. *Nature* **432**, 231–235.

Faller, M., Matsunaga, M., Yin, S., Loo, J. A., and Guo, F. (2007). Heme is involved in microRNA processing. *Nat. Struct. Mol. Biol.* **14**, 23–29.

Filippov, V., Solovyev, V., Filippova, M., and Gill, S. S. (2000). A novel type of RNase III family proteins in eukaryotes. *Gene* **245**, 213–221.

Fortin, K. R., Nicholson, R. H., and Nicholson, A. W. (2002). Mouse ribonuclease III. cDNA structure, expression analysis, and chromosomal location. *BMC Genomics* **3**, 26.

Goldberg, R., Motzkin, B., Marion, R., Scambler, P. J., and Shprintzen, R. J. (1993). Velo-cardio-facial syndrome: A review of 120 patients. *Am. J. Med. Genet.* **45**, 313–319.

Gregory, R. I., Yan, K. P., Amuthan, G., Chendrimada, T., Doratotaj, B., Cooch, N., and Shiekhattar, R. (2004). The Microprocessor complex mediates the genesis of micro-RNAs. *Nature* **432**, 235–240.

Grishok, A., Pasquinelli, A. E., Conte, D., Li, N., Parrish, S., Ha, I., Baillie, D. L., Fire, A., Ruvkun, G., and Mello, C. C. (2001). Genes and mechanisms related to RNA interference regulate expression of the small temporal RNAs that control *C. elegans* developmental timing. *Cell* **106**, 23–34.

Han, J., Lee, Y., Yeom, K. H., Kim, Y. K., Jin, H., and Kim, V. N. (2004). The Drosha-DGCR8 complex in primary microRNA processing. *Genes Dev.* **18**, 3016–3027.

Han, J., Lee, Y., Yeom, K. H., Nam, J. W., Heo, I., Rhee, J. K., Sohn, S. Y., Cho, Y., Zhang, B. T., and Kim, V. N. (2006). Molecular basis for the recognition of primary microRNAs by the Drosha-DGCR8 complex. *Cell* **125**, 887–901.

Hutvagner, G., McLachlan, J., Pasquinelli, A. E., Balint, E., Tuschl, T., and Zamore, P. D. (2001). A cellular function for the RNA-interference enzyme Dicer in the maturation of the let-7 small temporal RNA. *Science* **293**, 834–838.

Ketting, R. F., Fischer, S. E., Bernstein, E., Sijen, T., Hannon, G. J., and Plasterk, R. H. (2001). Dicer functions in RNA interference and in synthesis of small RNA involved in developmental timing in *C. elegans*. *Genes Dev.* **15**, 2654–2659.

Kim, V. N. (2005). MicroRNA biogenesis: coordinated cropping and dicing. *Nat. Rev. Mol. Cell Biol.* **6,** 376–385.

Khvorova, A., Reynolds, A., and Jayasena, S. D. (2003). Functional siRNAs and miRNAs exhibit strand bias. *Cell* **115,** 209–216.

Kloosterman, W. P., and Plasterk, R. H. (2006). The diverse functions of microRNAs in animal development and disease. *Dev. Cell* **11,** 441–450.

Knight, S. W., and Bass, B. L. (2001). A role for the RNase III enzyme DCR-1 in RNA interference and germ line development in *Caenorhabditis elegans. Science* **293,** 2269–2271.

Kurihara, Y., and Watanabe, Y. (2004). Arabidopsis micro-RNA biogenesis through Dicer-like 1 protein functions. *Proc. Natl. Acad. Sci. USA* **101,** 12753–12758.

Landthaler, M., Yalcin, A., and Tuschl, T. (2004). The human DiGeorge syndrome critical region gene 8 and its *D. melanogaster* homolog are required for miRNA biogenesis. *Curr. Biol.* **14,** 2162–2167.

Lee, Y., Kim, M., Han, J., Yeom, K. H., Lee, S., Baek, S. H., and Kim, V. N. (2004). MicroRNA genes are transcribed by RNA polymerase II. *EMBO J.* **23,** 4051–4060.

Lee, Y., Ahn, C., Han, J., Choi, H., Kim, J., Yim, J., Lee, J., Provost, P., Radmark, O., Kim, S., and Kim, V. N. (2003). The nuclear RNase III Drosha initiates microRNA processing. *Nature* **425,** 415–419.

Lund, E., Guttinger, S., Calado, A., Dahlberg, J. E., and Kutay, U. (2004). Nuclear export of microRNA precursors. *Science* **303,** 95–98.

Schwarz, D. S., *et al.* (2003). Asymmetry in the assembly of the RNAi enzyme complex. *Cell* **115,** 199–208.

Shiohama, A., Sasaki, T., Noda, S., Minoshima, S., and Shimizu, N. (2003). Molecular cloning and expression analysis of a novel gene DGCR8 located in the DiGeorge syndrome chromosomal region. *Biochem. Biophys. Res. Commun.* **304,** 184–190.

Wang, Y., Medvid, R., Melton, C., Jaenisch, R., and Blelloch, R. (2007). DGCR8 is essential for microRNA biogenesis and silencing of embryonic stem cell self-renewal. *Nat. Genet.* **39,** 380–385.

Wu, H., Xu, H., Miraglia, L. J., and Crooke, S. T. (2000). Human RNase III is a 160-kDa protein involved in preribosomal RNA processing. *J. Biol. Chem.* **275,** 36957–36965.

Yi, R., Qin, Y., Macara, I. G., and Cullen, B. R. (2003). Exportin-5 mediates the nuclear export of pre-microRNAs and short hairpin RNAs. *Genes Dev.* **17,** 3011–3016.

Zeng, Y., and Cullen, B. R. (2005). Efficient processing of primary microRNA hairpins by Drosha requires flanking nonstructured RNA sequences. *J. Biol. Chem.* **280,** 27595–27603.

Zeng, Y., Yi, R., and Cullen, B. R. (2005). Recognition and cleavage of primary micro-RNA precursors by the nuclear processing enzyme Drosha. *EMBO J.* **24,** 138–148.

CHAPTER SIX

MICROARRAY ANALYSIS OF MIRNA GENE EXPRESSION

J. Michael Thomson,* Joel S. Parker,† *and* Scott M. Hammond*

Contents

Abstract

MicroRNAs (miRNAs) are small, noncoding RNAs that regulate the expression of target mRNAs. Although thousands of miRNAs have been identified, few have been functionally linked to specific biological pathways. Microarray-based expression analysis is an ideal strategy for identifying candidate miRNAs that correlate with biological pathways and for generating molecular signatures of disease states. This chapter will describe a simple, low-cost microarray platform optimized for miRNA expression analysis.

1. INTRODUCTION

miRNAs were discovered over 20 years ago in the nematode *Caenorhabditis elegans* (Lee *et al.*, 1993; Wightman *et al.*, 1993). This founding miRNA, *lin-4*, was essential for proper timing during larval development.

* Department of Cell and Developmental Biology, University of North Carolina, Chapel Hill, North Carolina
† Constella Group, Durham, North Carolina

Methods in Enzymology, Volume 427
ISSN 0076-6879, DOI: 10.1016/S0076-6879(07)27006-5

The *lin-4* gene did not code for a protein. Rather, it generated a 22 nucleotide RNA that negatively regulated the translational efficiency of the mRNA for *lin-14*. This groundbreaking discovery led to a rapidly growing field in biological research. Over 4000 miRNAs have been identified across 49 genomes (Griffiths-Jones, 2006). While it is clear that miRNA regulation is a critical component of many biological pathways, most known miRNA genes have not been linked to specific pathways.

Microarray analysis of gene expression is a well-established method to identify genes that correlate with cellular processes. This approach has been adapted for miRNA genes. Initial efforts sought to catalog the miRNA transcriptome across adult and embryonic tissues (Babak *et al.*, 2004; Barad *et al.*, 2004; Baskerville and Bartel, 2005; Castoldi *et al.*, 2006; Liu *et al.*, 2004a; Miska *et al.*, 2004; Neilson *et al.*, 2007; Nelson *et al.*, 2004; Shingara *et al.*, 2005; Sun *et al.*, 2004; Thomson *et al.*, 2004). As expected, many miRNAs exhibit tissue-restricted expression. In several cases, this restricted expression has been linked to function. For example, miR-1 and miR-133 are highly restricted to cardiac and skeletal muscle, and have been shown to be important for proper development of these tissues (Boutz *et al.*, 2007; Chen *et al.*, 2006; Nakajima *et al.*, 2006; Rao *et al.*, 2006; Sokol and Ambros, 2005; Zhao *et al.*, 2005).

The use of mRNA microarrays for clinical diagnostics has been under development for several years, and such a device has received Food and Drug Administration (FDA) approval for breast cancer diagnosis (Glas *et al.*, 2006). Following this lead, cancer researchers have explored the use of miRNA expression data for diagnostics. The small size of the mature miRNA is less susceptible to nuclease degradation. Additionally, the small size makes it possible to extract miRNAs from paraffin-embedded formalin-fixed tissue blocks, which makes large archives of fixed tissue available for molecular analysis (Nelson *et al.*, 2006).

These factors have encouraged numerous studies that have examined miRNA expression patterns in normal tissue versus tumor tissue and cell lines. miRNA signatures have emerged for several tumor types. For example, one study examined miRNA expression levels for primary hepatocellular carcinoma compared to normal adjacent tissue (Murakami *et al.*, 2006). The cancerous tissue had a clear miRNA signature. Several miRNAs were consistently overexpressed, including miR-18 and miR-224, whereas many miRNAs had decreased expression. Distinct signatures were also apparent for poorly differentiated tumors compared to more highly differentiated tumors, and for chronic hepatitis compared to cirrhosis. Using these signatures the authors developed a miRNA classifier algorithm that accurately predicted the clinical state of liver tissue.

Several other microarray studies have demonstrated altered miRNA expression in glioblastoma, chronic lymphocytic leukemia, Burkitt's lymphoma, colorectal, lung, and breast carcinoma, and others (Cummins and Velculescu, 2006).

We were one of the first labs to develop a custom miRNA microarray platform (Thomson *et al.*, 2004). We have optimized labeling and hybridization procedures for the unique features of miRNAs. Using this platform, we have performed more than 3000 microarray hybridizations and have published numerous research papers on findings derived from this (Chen *et al.*, 2006; Giraldez *et al.*, 2005; He *et al.*, 2005; Liu *et al.*, 2004b; Perkins *et al.*, 2007; Poliseno *et al.*, 2006). One factor we stressed was minimizing the cost of the platform, which makes it attractive compared to commercially available platforms. This chapter discusses methodologies for developing and utilizing a custom miRNA microarray.

2. OVERVIEW OF miRNA BIOGENESIS AND EFFECTOR PATHWAYS

miRNA-mediated gene repression begins with the transcription of a *primary transcript,* also termed the *pri-miRNA* (Fig. 6.1). The miRNA stem-loop itself can be located within an intron or an exon of the transcript. Following transcription, the pri-miRNA is processed sequentially by two ribonuclease III enzymes (RNase III) (Kim, 2005). The first enzyme,

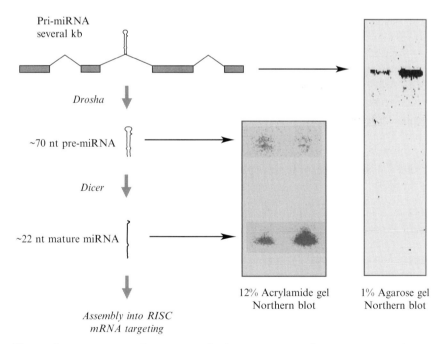

Figure 6.1 Biogenesis of miRNAs. The biogenesis steps for miRNAs are shown. After the mature miRNA species is produced, it is incorporated into the RNA-induced silencing complex (RISC) where it directs repression of targeted mRNAs (not shown). Northern blot detection of each intermediate is shown for illustrative purposes. (See color insert.)

Drosha, excises the stem-loop structure from the pri-miRNA. The stem-loop, termed the *precursor* or *pre-miRNA*, is exported to the cytoplasm where it is a substrate for the RNase III enzyme, Dicer. The product of this reaction is an RNA species of approximately 22 nucleotides in length and is referred to as a *mature* miRNA.

The mature miRNA is loaded into the RNA-induced silencing complex (RISC) (Du and Zamore, 2005). The RISC locates mRNA sequence elements that are complementary to the miRNA. If the miRNA and the mRNA base pairing is perfect in the central region of the miRNA, the mRNA is endonucleolytically cleaved by Argonaute (Liu *et al.*, 2004b). The resultant mRNA fragments are rapidly degraded, leading to efficient gene repression. If the miRNA–mRNA base pairing is imperfect, the mRNA is not cleaved. Rather, the translational efficiency of the mRNA is reduced. The exact mechanism of translational suppression is not clear, as several competing models have been proposed (Humphreys *et al.*, 2005; Liu *et al.*, 2005; Maroney *et al.*, 2006; Nottrott *et al.*, 2006; Petersen *et al.*, 2006; Pillai *et al.*, 2005; Sen and Blau, 2005). The end result is decreased translation of the targeted mRNA.

3. miRNA Expression Analysis Strategies

Early work in the field relied on Northern blot analysis for measurement of miRNA expression. All three biogenesis intermediates are amenable to Northern analysis (see Fig. 6.1). Because the mature miRNA is the only species that is biologically active, it is this species that provides the most information about function. It was originally believed that miRNA biogenesis steps were not regulated. That is, the mature miRNA levels would correlate with the pri-miRNA and pre-miRNA. However, we have reported that many miRNAs are regulated at the Drosha processing step (Thomson *et al.*, 2006). Regulation at the Dicer processing step has also been reported (Obernosterer *et al.*, 2006; Wulczyn *et al.*, 2007). For this reason, it is unsafe to assume that the expression of pri-miRNAs or pre-miRNAs will correlate with the more relevant mature miRNA. Therefore, we recommend expression analysis based on the mature miRNA species.

Another way to quantitate mature miRNA expression is with a modified quantitative reverse transcription PCR (qRT-PCR). Because the mature miRNA lacks a poly-A tail, it is not possible to use oligo-dT–directed reverse transcription. Chiang and Shi have developed a protocol to overcome this (Shi and Chiang, 2005). An artificial tail is added to the miRNA with *Escherichia coli* poly-A polymerase (PAP). This permits the use of an anchored oligo dT primer for the reverse transcription step. PCR is performed with primers directed against the mature miRNA and against a common sequence contained within the reverse transcription primer. An alternative protocol, based on Taqman PCR, is available from Applied Biosystems (Foster City, CA)(Chen *et al.*, 2005). This protocol has higher sensitivity but requires the kit with the dual modified oligonucleotide.

High-throughput expression analysis is possible with microarrays (described later) or with a novel flow cytometric assay available from Luminex (Lu *et al.*, 2005). In this assay, monodisperse beads are conjugated with antisense probes to individual miRNAs. Each set of beads is labeled with a unique fluorescent color which provides the identity of the miRNA probe. After hybridization with miRNA samples, the beads are analyzed with a specialized flow cytometer. One hundred miRNAs can be assayed in a single experiment. This method has been proven to be highly specific because the hybridization occurs in solution. This approach was used to profile the miRNA expression in 334 human tumors and normal tissues, providing a comprehensive molecular signature for many cancer types (Lu *et al.*, 2005).

4. Considerations for miRNA Microarrays

Several features of mature miRNAs present problems for conventional microarray platforms. First, the miRNA cannot be labeled with a protocol that begins with oligo dT reverse transcription. Because the $3'$ end of the mature miRNA contains a hydroxyl group, several enzymatic reactions can be used to append labeled nucleic acids. Terminal deoxynucleotide transferase (TdT) and PAP will tail an RNA with a $3'$ hydroxyl group. Inclusion of fluorescent conjugated nucleotides permits direct labeling. The RNA can be indirectly labeled using biotin conjugated nucleotides, followed by streptavidin labeling, or by inclusion of amine-modified nucleotides, followed by N-hydroxylsuccinimide ester (NHS) labeling. In general, these methods may lead to variability in labeling because the number of fluorescent moieties depends on the tail length. We have developed a direct labeling strategy that uses T4 RNA Ligase to link a single fluorescent Cy3 to the $3'$ end of the miRNA (Igloi, 1996; Thomson *et al.*, 2004). This method has been refined and is the recommended protocol for Agilent miRNA microarrays (Wang *et al.*, 2007a). The method very efficiently labels RNA and is easy to perform. Total RNA is used, eliminating the need to fractionate the miRNA population. Minimal postlabeling cleanup is required. This is our recommended protocol and is described in detail later. Note: Most plant miRNAs are methylated at the $2'$ position of the $3'$ end of the miRNA (Yang *et al.*, 2006; Yu *et al.*, 2005). This moiety reduces efficiency of ligation and PAP-based labeling methods.

The second major consideration for miRNA microarrays is the short nature of the miRNA probes. Because the entire mature miRNA is only ~22 nucleotides, it is not possible to adjust the probe sequence to normalize melting temperature. Several commercial platforms have devised solutions to this. Agilent probes have a stem-loop sequence attached to the miRNA complementary region. When the miRNA hybridizes, it can base-stack with the duplex region of the stem. This increases base pairing energy, allowing more stringent hybridization conditions (Wang *et al.*, 2007a). Exiqon probes, in contrast, are designed with locked nucleic acid (LNA) bases interspersed within the DNA sequence (Castoldi *et al.*, 2006). This affects the helical structure of the hybrid and increases base pairing energy. Again, higher stringency conditions are possible. Our platform is based on robotic spotted microarrays and thus is compatible with a standard DNA probe set, or an LNA probe set, which is available from Exiqon (Woburn, MA).

5. Data Analysis and Interpretation

Microarray data processing includes processing of the scanned image and analysis of processed data. Many technologies currently used for mRNA microarrays are directly applicable to miRNA arrays. Guidelines for experimental design, quantification of the image, background estimation, normalization, and testing for differential abundance have emerged (Allison *et al.*, 2006). However, there are specific characteristics of a miRNA array experiment that should be considered when choosing the appropriate method of analysis.

Normalization in mRNA microarray experiments is performed to adjust for variation not due to the experiment. In contrast with mRNA microarray normalization, convergence to the optimal methods for miRNA arrays has not yet occurred. Some have chosen to bypass normalization, which may be adequate after much investigation of the positive and negative controls along with spike-in intensities (Barad *et al.*, 2004; Baskerville and Bartel, 2005; Liang *et al.*, 2005). However, when normalization is performed properly, nonexperimental variance is reduced, allowing for more accurate identification of results. Positive controls (Lu *et al.*, 2005; Wang *et al.*, 2007b), linear adjustment (Miska *et al.*, 2004; Thomson *et al.*, 2004), and forms of lowess (Babak *et al.*, 2004; Perkins *et al.*, 2007), have all been utilized in published work.

The normalization methods often utilized, such as quantile and intensity-dependent lowess, perform well for large-scale mRNA data, but assume some level of symmetry in differential expression. As has been shown, this assumption may be incorrect due to the global change seen in certain phenotypes (Lu *et al.*, 2005; Wang *et al.*, 2007b). Such results highlight the need of proper normalization because a method with the symmetry assumption will hide a global effect.

Choices for normalization exist that do not assume symmetry or are robust to relatively large deviations. Normalization to spike in controls, housekeeping genes, or spots representing a pool of transcripts does not assume symmetry, while variance-stabilized normalization and rank-invariant normalization were developed to be less sensitive to these deviations (Huber *et al.*, 2002; Li and Hung Wong, 2001). A method has been proposed (wlowess) that combines lowess with a pool of sample transcripts (Oshlack *et al.*, 2007). None of these methods have shown to be optimal in all cases, and the optimal normalization must be selected using experiment-specific knowledge.

Statistical inference of differential abundance between two classes of samples can better capitalize on the methods from mRNA arrays. One problem of identifying differential expression in mRNA array data is the large number of measurements relative to the number of samples. A second barrier is the multiple testing problem. An miRNA array generally assays a few hundred transcripts instead of tens of thousands, but this is still enough tests to report false positives when traditional *p*-value thresholds are used. Significance Analysis of Microarray (SAM) and LIMMA are examples of popular statistical methods that attempt to overcome these problems (Smyth, 2004; Tusher *et al.*, 2001).

6. Validation Strategies

After the microarray data has been analyzed, the final step is validation. In general, strategies applicable for mRNA microarrays can be used for miRNA microarrays, with modifications. Validation of expression differences can be performed by qRT-PCR using the methods described previously (Chen *et al.*, 2005; Shi and Chiang, 2005). If the microarray is being used as a gene-finding tool, functional validation strengthens the interpretation of data. This would include miRNA overexpression or inhibition, followed by the appropriate functional assay. This is highly dependent on the biological question and cannot be addressed here.

Finally, the expression data should be archived in a standardized database. Both NCBI-GEO (http://www.ncbi.nlm.nih.gov/geo/) and ArrayExpress (http://www.ebi.ac.uk/arrayexpress/) databases have been used for storage of miRNA microarray data.

7. miRNA Microarray Protocol

7.1. Microarray spotting

Oligonucleotides probes complementary to the mature miRNA were synthesized and spotted (40 μM) on Corning GAPS-2 slides in $3\times$ SSC using an appropriate robotic spotter. Alternatively, commercial probe sets composed of DNA or DNA/LNA (locked nucleic acid) oligonucleotides can be printed. After printing, the slides are stored under nitrogen gas to protect them from oxidation.

7.2. RNA isolation

Isolation of total RNA is done by the phenolic separation of DNA and RNA using Trizol (Invitrogen, Carlsbad, CA). It is important not to use other procedures, that is, kits that remove low-molecular-weight RNA. However, kits that specifically isolate the low-molecular-weight RNA species work fine (i.e., mirVANA, Ambion, Austin, TX). Trizol-extracted RNA is sufficiently free of contaminating DNA as not to affect the concentration of the nucleic acids measured. The existence of small amounts of DNA does not inhibit the labeling or hybridization. Isolated RNA is suspended in diethyl-pyrocarbonate (DEPC)–treated water, preferably at a concentration of 1 μg/μl.

7.3. Labeling and cleanup

Mature miRNAs are phosphorylated on the 5$'$ end and contain a 3$'$ hydroxyl moiety. The labeling procedure exploits the 3$'$ hydroxyl end by T4 RNA ligase-mediated ligation of a p-CU-Cy3 dinucleotide (TriLink BioTech, San Diego, CA). Generally, 5 μg is labeled per array, though as little as 500 ng gives good hybridization signal.

RNA sample (0.1–1 μg/μl)   6 μl
p-CU-Cy3 dinucleotide 250 ng/μl   1 μl
10× Igloi buffer   2 μl
DMSO 100% (25% final)   5 μl
T4 RNA Ligase 20 units (NEB)   1 μl
DEPC dH$_2$O   5 μl
Final volume: 20 μl

10× Igloi buffer:
1 mM ATP
500 mM HEPES, pH 7.8
35 mM DTT
200 mM MgCl$_2$
10 μg/ml BSA

1. To ensure normalized sample labeling, make a master mix of the (Cy3 dinucleotide, 10× buffer, and DMSO). All ligations are performed in amber microfuge tubes to protect the Cy3 from photobleaching.
2. Mix the RNA and 13 μl of the master mix, then add the T4 RNA ligase and mix completely. This ensures the miRNAs are saturated with labeled Cy3 dinucleotide to avoid circularization or concatenation of the miRNA.

3. Centrifuge briefly to collect the sample in the bottom of the tube and place on ice for 1 to 2 h.

4. Make a master mix for precipitation of the labeling reaction with 10 to 20% extra. Add 46 μl of water/sodium acetate mix to each tube. Mix the samples and add 150 μl 100% ethanol and vortex. Place on ice for 10 min and centrifuge for 10 min. Carefully aspirate the supernatant using a gel-loading tip and add 150 μl of 70% ethanol. Mix gently and centrifuge for 5 min. Remove supernatant again with a gel-loading tip being very careful not to disturb the pellet. The pellet should be visible as a red dot. To array the sample, dissolve the pellets in 6 μl of DEPC dH$_2$O. Use heat at 65° for about 3 to 5 min with intermittent vortexing.

Water/sodium acetate mix
DEPC-treated dH$_2$O	40 μl
3-M sodium acetate	5 μl
Glycogen 20 mg/ml	1 μl
Final volume:	46 μl

7.4. Hybridization

1. During the labeling procedure, prepare the array slides. Wear gloves! First, mark the grid corners by etching on the back side of the slides with a diamond knife (Fig. 6.2). Be careful to only touch the side of the slides avoiding the arrays. Slides are then cross-linked with UV at 600 mJ (Stratalinker, Stratagene, La Jolla, CA).

2. When complete, remove the slides and place in a slide holder. Add enough blocking buffer to cover the slide and put in the hybridization oven at 37° for at least 1 h.

3. Pour off the blocking buffer and wash 3 to 4 times with DEPC-treated dH$_2$O. A final wash with 100% ethanol is done to speed up slide drying. Allow slide to dry completely.

4. Hybridization is performed in Frame-Seal (Bio-Rad, Hercules, CA) chambers. Set frames over grids by peeling off the open side of the frame seal and placing it over the grid (Fig. 6.3). Peel back the other side and place a gel-loading tip in the corner, avoiding the grid. Replace the cover over the entire frame seal and gel-loading tip. Seal the frame up to the tip, leaving about 6 to 10 mM unsealed from the tip (see Fig. 6.3). This allows for good displacement of air during the filling process. Pre-heat the slide at 37° prior to loading the chamber.

5. Add 60 μl of Church and Gilbert's hybridization buffer to the solubilized labeled RNA. Vortex briefly and centrifuge. Heat samples for 3 to 4 min at 95°. Remove from the heat and immediately load sample into the

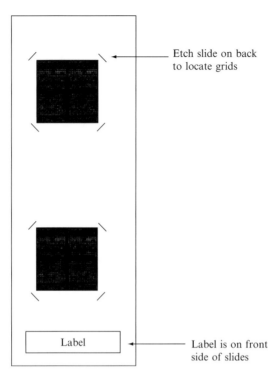

Etch slide on back
to locate grids

Label

Label is on front
side of slides

Figure 6.2   Appearance of microRNA microarray slide. (See color insert.)

gel-loading tip using a 200-μl tip. Be gentle here. Load sample in one
clean motion, and allow capillary action to take over. Once the sample is
loaded, remove gel loading tip and press coverslip down to seal. Invert
slide with a few brisk motions to remove any hydrophobic pockets that
may have formed. Bubbles are acceptable, as the slides will rotate and the
bubbles will facilitate a good mixing process. Place the slides in a rotating
hybridization oven and rotate on slowest setting 1 to 2 h at 37°.

Church and Gilbert's hybridization buffer:
400 mM Na$_2$HPO$_4$, pH 7.0
0.8% BSA
5% SDS
12% formamide

Blocking buffer:
3× SSC
0.1% SDS
0.2% BSA

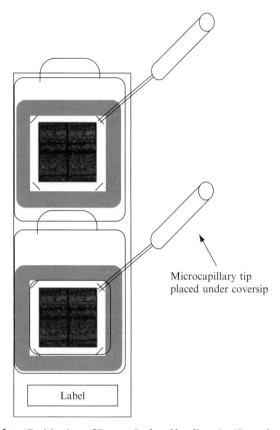

Microcapillary tip
placed under coversip

Label

Figure 6.3   Positioning of Frame-Seal and loading tip. (See color insert.)

Washing

2× SSC/0.025% SDS @ room temp
0.8× SSC @ 23°, performed three times
0.4× SSC @ ice cold, performed two times

Remove the slides from the cassette and place directly into a wash container holding the 2× SSC/0.025% SDS solution. Dunk the slides up and down several times gently and leave for 3 min, then transfer to a container with 0.8× SSC and wash 3 min with intermittent gentle dunking. Wash again in 0.8× SSC two more times for 3 min, then two washes of ice-cold 0.4× SSC. The last two washes are done in ice-cold buffer. Chilling slides prior to centrifugation removes the problems associated with salt. The slides are then transferred to a swinging bucket rotor and centrifuged 10 min at 500 rpm (100×g). Be sure to make the transfer to the centrifuge quick. You want the slides to remain cold.

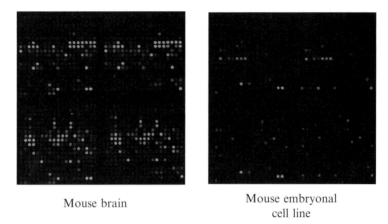

Mouse brain

Mouse embryonal
cell line

Figure 6.4   Images from scanned microarrays. Scans from two microarray experiments are shown for illustrative purposes. (See color insert.)

Scan the slides in the 532-nM channel on an appropriate microarray scanner. Figure 6.4 shows scanned images from two microarray experiments. The brain sample has high expression of many miRNAs. The P19 teratocarcinoma cell line, in contrast, has very low expression of most miRNAs, with the exception of a small number of stem cell miRNAs.

8. DATA ANALYSIS

To demonstrate the natural progression of analysis, we refer to a study published by Perkins *et al.* (2007). Microarray data analysis began with data extraction from the GPR files. Data points were eliminated if foreground was not 1.5 times local background and a probe was removed if more than 40% of the data points were missing. These thresholds are chosen based on overall array intensities and analysis of negative control intensities when available. A total of 239 miRNA remained after this preprocessing.

Data were background-subtracted and log-transformed, and missing values were inputted using k-NN (Troyanskaya *et al.*, 2001). For comparisons across samples, data were normalized using rank invariant normalization (Li and Hung Wong, 2001). The rank invariant procedure was chosen due to its robustness to asymmetry in intensity distribution. The per-sample mean of the two rank invariant normalized probes was used for analyses.

This experiment consisted of two experimental groups without matching of cases and controls. Therefore, univariate calculations of differential expression were estimated using Statistical SAM (two-class, unpaired test; 500 permutations; FDR of 5%) (Tusher *et al.*, 2001). All the aforementioned algorithms are available in R via Bioconductor and other packages. Finally, GeneCluster and TreeView were utilized to visually inspect the results.

ACKNOWLEDGMENTS

The authors acknowledge Charles Perou and Jason Lieb for assistance with design and production of microarrays. The authors thank members of the Hammond lab for comments on the manuscript. Support for work is from the American Association for Cancer Research, US Army Congressional Directed Medical Research Program, National Institutes of Health, and American Cancer Society.

REFERENCES

Allison, D. B., Cui, X., Page, G. P., and Sabripour, M. (2006). Microarray data analysis: From disarray to consolidation and consensus. *Nat. Rev. Genet.* **7,** 55–65.

Babak, T., Zhang, W., Morris, Q., Blencowe, B. J., and Hughes, T. R. (2004). Probing microRNAs with microarrays: Tissue specificity and functional inference. *RNA* **10,** 1813–1819.

Barad, O., Meiri, E., Avniel, A., Aharonov, R., Barzilai, A., Bentwich, I., Einav, U., Gilad, S., Hurban, P., Karov, Y., Lobenhofer, E. K., Sharon, E., *et al.* (2004). MicroRNA expression detected by oligonucleotide microarrays: System establishment and expression profiling in human tissues. *Genome Res.* **14,** 2486–2494.

Baskerville, S., and Bartel, D. P. (2005). Microarray profiling of microRNAs reveals frequent coexpression with neighboring miRNAs and host genes. *RNA* **11,** 241–247.

Boutz, P. L., Chawla, G., Stoilov, P., and Black, D. L. (2007). MicroRNAs regulate the expression of the alternative splicing factor nPTB during muscle development. *Genes Dev.* **21,** 71–84.

Castoldi, M., Schmidt, S., Benes, V., Noerholm, M., Kulozik, A. E., Hentze, M. W., and Muckenthaler, M. U. (2006). A sensitive array for microRNA expression profiling (miChip) based on locked nucleic acids (LNA). *RNA* **12,** 913–920.

Chen, C., Ridzon, D. A., Broomer, A. J., Zhou, Z., Lee, D. H., Nguyen, J. T., Barbisin, M., Xu, N. L., Mahuvakar, V. R., Andersen, M. R., Lao, K. Q., Livak, K. J., *et al.* (2005). Real-time quantification of microRNAs by stem-loop RT-PCR. *Nucleic Acids Res.* **33,** e179.

Chen, J. F., Mandel, E. M., Thomson, J. M., Wu, Q., Callis, T. E., Hammond, S. M., Conlon, F. L., and Wang, D. Z. (2006). The role of microRNA-1 and microRNA-133 in skeletal muscle proliferation and differentiation. *Nat. Genet.* **38,** 228–233.

Cummins, J. M., and Velculescu, V. E. (2006). Implications of micro-RNA profiling for cancer diagnosis. *Oncogene* **25,** 6220–6227.

Du, T., and Zamore, P. D. (2005). microPrimer: The biogenesis and function of micro-RNA. *Development* **132,** 4645–4652.

Giraldez, A. J., Cinalli, R. M., Glasner, M. E., Enright, A. J., Thomson, J. M., Baskerville, S., Hammond, S. M., Bartel, D. P., and Schier, A. F. (2005). MicroRNAs regulate brain morphogenesis in zebrafish. *Science* **308,** 833–838.

Glas, A. M., Floore, A., Delahaye, L. J., Witteveen, A. T., Pover, R. C., Bakx, N., Lahti-Domenici, J. S., Bruinsma, T. J., Warmoes, M. O., Bernards, R., Wessels, L. F., and Van't Veer, L. J. (2006). Converting a breast cancer microarray signature into a high-throughput diagnostic test. *BMC Genomics* **7,** 278.

Griffiths-Jones, S. (2006). miRBase: The microRNA sequence database. *Methods Mol. Biol.* **342,** 129–138.

He, L., Thomson, J. M., Hemann, M. T., Hernando-Monge, E., Mu, D., Goodson, S., Powers, S., Cordon-Cardo, C., Lowe, S. W., Hannon, G. J., and Hammond, S. M. (2005). A microRNA polycistron as a potential human oncogene. *Nature* **435,** 828–833.

Huber, W., von Heydebreck, A., Sultmann, H., Poustka, A., and Vingron, M. (2002). Variance stabilization applied to microarray data calibration and to the quantification of differential expression. *Bioinformatics* **18**(Suppl. 1), S96–S104.

Humphreys, D. T., Westman, B. J., Martin, D. I., and Preiss, T. (2005). MicroRNAs control translation initiation by inhibiting eukaryotic initiation factor 4E/cap and poly (A) tail function. *Proc. Natl. Acad. Sci. USA* **102,** 16961–16966.

Igloi, G. L. (1996). Nonradioactive labeling of RNA. *Anal. Biochem.* **233,** 124–129.

Kim, V. N. (2005). MicroRNA biogenesis: Coordinated cropping and dicing. *Nat. Rev. Mol. Cell. Biol.* **6,** 376–385.

Lee, R. C., Feinbaum, R. L., and Ambros, V. (1993). The *C. elegans* heterochronic gene *lin-4* encodes small RNAs with antisense complementarity to *lin-14. Cell* **75,** 843–854.

Li, C., and Hung, Wong W. (2001). Model-based analysis of oligonucleotide arrays: Model validation, design issues and standard error application. *Genome Biol.* **2,** RESEARCH0032.

Liang, R. Q., Li, W., Li, Y., Tan, C. Y., Li, J. X., Jin, Y. X., and Ruan, K. C. (2005). An oligonucleotide microarray for microRNA expression analysis based on labeling RNA with quantum dot and nanogold probe. *Nucleic Acids Res.* **33,** e17.

Liu, C. G., Calin, G. A., Meloon, B., Gamliel, N., Sevignani, C., Ferracin, M., Dumitru, C. D., Shimizu, M., Zupo, S., Dono, M., Alder, H., Bullrich, F., Negrini, M., and Croce, C. M. (2004a). An oligonucleotide microchip for genome-wide microRNA profiling in human and mouse tissues. *Proc. Natl. Acad. Sci. USA* **101,** 9740–9744.

Liu, J., Carmell, M. A., Rivas, F. V., Marsden, C. G., Thomson, J. M., Song, J. J., Hammond, S. M., Joshua-Tor, L., and Hannon, G. J. (2004b). Argonaute2 is the catalytic engine of mammalian RNAi. *Science* **305,** 1437–1441.

Liu, J., Valencia-Sanchez, M. A., Hannon, G. J., and Parker, R. (2005). MicroRNA-dependent localization of targeted mRNAs to mammalian P-bodies. *Nat. Cell Biol.* **7,** 719–723.

Lu, J., Getz, G., Miska, E. A., Alvarez-Saavedra, E., Lamb, J., Peck, D., Sweet-Cordero, A., Ebert, B. L., Mak, R. H., Ferrando, A. A., Downing, J. R., Jacks, T., *et al.* (2005). MicroRNA expression profiles classify human cancers. *Nature* **435,** 834–838.

Maroney, P. A., Yu, Y., Fisher, J., and Nilsen, T. W. (2006). Evidence that microRNAs are associated with translating messenger RNAs in human cells. *Nat. Struct. Mol. Biol.* **13,** 1102–1107.

Miska, E. A., Alvarez-Saavedra, E., Townsend, M., Yoshii, A., Sestan, N., Rakic, P., Constantine-Paton, M., and Horvitz, H. R. (2004). Microarray analysis of microRNA expression in the developing mammalian brain. *Genome Biol.* **5,** R68.

Murakami, Y., Yasuda, T., Saigo, K., Urashima, T., Toyoda, H., Okanoue, T., and Shimotohno, K. (2006). Comprehensive analysis of microRNA expression patterns in hepatocellular carcinoma and non-tumorous tissues. *Oncogene* **25,** 2537–2545.

Nakajima, N., Takahashi, T., Kitamura, R., Isodono, K., Asada, S., Ueyama, T., Matsubara, H., and Oh, H. (2006). MicroRNA-1 facilitates skeletal myogenic differentiation without affecting osteoblastic and adipogenic differentiation. *Biochem. Biophys. Res. Commun.* **350,** 1006–1012.

Neilson, J. R., Zheng, G. X., Burge, C. B., and Sharp, P. A. (2007). Dynamic regulation of miRNA expression in ordered stages of cellular development. *Genes Dev.* **21,** 578–589.

Nelson, P. T., Baldwin, D. A., Kloosterman, W. P., Kauppinen, S., Plasterk, R. H., and Mourelatos, Z. (2006). RAKE and LNA-ISH reveal microRNA expression and localization in archival human brain. *RNA* **12,** 187–191.

Nelson, P. T., Baldwin, D. A., Scearce, L. M., Oberholtzer, J. C., Tobias, J. W., and Mourelatos, Z. (2004). Microarray-based, high-throughput gene expression profiling of microRNAs. *Nat. Methods* **1,** 155–161.

Nottrott, S., Simard, M. J., and Richter, J. D. (2006). Human *let-7a* miRNA blocks protein production on actively translating polyribosomes. *Nat. Struct. Mol. Biol.* **13,** 1108–1114.

Obernosterer, G., Leuschner, P. J., Alenius, M., and Martinez, J. (2006). Post-transcriptional regulation of microRNA expression. *RNA* **12,** 1161–1167.

Oshlack, A., Emslie, D., Corcoran, L. M., and Smyth, G. K. (2007). Normalization of boutique two-color microarrays with a high proportion of differentially expressed probes. *Genome Biol.* **8,** R2.

Perkins, D. O., Jeffries, C., Jarskog, L. F., Thomson, J. M., Woods, K., Newman, M. A., Parker, J. S., Jin, J., and Hammond, S. M. (2007). miRNA expression in the prefrontal cortex of individuals with schizophrenia and schizoaffective disorder. *Genome Biol.* **8,** R27.

Petersen, C. P., Bordeleau, M. E., Pelletier, J., and Sharp, P. A. (2006). Short RNAs repress translation after in itiation in mammalian cells. *Mol. Cell* **21,** 533–542.

Pillai, R. S., Bhattacharyya, S. N., Artus, C. G., Zoller, T., Cougot, N., Basyuk, E., Bertrand, E., and Filipowicz, W. (2005). Inhibition of translational initiation by Let-7 MicroRNA in human cells. *Science* **309,** 1573–1576.

Poliseno, L., Tuccoli, A., Mariani, L., Evangelista, M., Citti, L., Woods, K., Mercatanti, A., Hammond, S., and Rainaldi, G. (2006). MicroRNAs modulate the angiogenic properties of HUVECs. *Blood* **108,** 3068–3071.

Rao, P. K., Kumar, R. M., Farkhondeh, M., Baskerville, S., and Lodish, H. F. (2006). Myogenic factors that regulate expression of muscle-specific microRNAs. *Proc. Natl. Acad. Sci. USA* **103,** 8721–8726.

Sen, G. L., and Blau, H. M. (2005). Argonaute 2/RISC resides in sites of mammalian mRNA decay known as cytoplasmic bodies. *Nat. Cell Biol.* **7,** 633–636.

Shi, R., and Chiang, V. L. (2005). Facile means for quantifying microRNA expression by real-time PCR. *Biotechniques* **39,** 519–525.

Shingara, J., Keiger, K., Shelton, J., Laosinchai-Wolf, W., Powers, P., Conrad, R., Brown, D., and Labourier, E. (2005). An optimized isolation and labeling platform for accurate microRNA expression profiling. *RNA* **11,** 1461–1470.

Smyth, G. K. (2004). Linear models and empirical bayes methods for assessing differential expression in microarray experiments. *Stat. Appl. Genet. Mol. Biol.* **3,** Article3.

Sokol, N. S., and Ambros, V. (2005). Mesodermally expressed *Drosophila* microRNA-1 is regulated by Twist and is required in muscles during larval growth. *Genes Dev.* **19,** 2343–2354.

Sun, Y., Koo, S., White, N., Peralta, E., Esau, C., Dean, N. M., and Perera, R. J. (2004). Development of a micro-array to detect human and mouse microRNAs and characterization of expression in human organs. *Nucleic Acids Res.* **32,** e188.

Thomson, J. M., Newman, M., Parker, J. S., Morin-Kensicki, E. M., Wright, T., and Hammond, S. M. (2006). Extensive post-transcriptional regulation of microRNAs and its implications for cancer. *Genes Dev.* **20,** 2202–2207.

Thomson, J. M., Parker, J., Perou, C. M., and Hammond, S. M. (2004). A custom microarray platform for analysis of microRNA gene expression. *Nat. Methods* **1,** 47–53.

Troyanskaya, O., Cantor, M., Sherlock, G., Brown, P., Hastie, T., Tibshirani, R., Botstein, D., and Altman, R. B. (2001). Missing value estimation methods for DNA microarrays. *Bioinformatics* **17,** 520–525.

Tusher, V. G., Tibshirani, R., and Chu, G. (2001). Significance analysis of microarrays applied to the ionizing radiation response. *Proc. Natl. Acad. Sci. USA* **98,** 5116–5121.

Wang, H., Ach, R. A., and Curry, B. (2007a). Direct and sensitive miRNA profiling from low-input total RNA. *RNA* **13,** 151–159.

Wang, Y., Medvid, R., Melton, C., Jaenisch, R., and Blelloch, R. (2007b). DGCR8 is essential for microRNA biogenesis and silencing of embryonic stem cell self-renewal. *Nat. Genet.* **39,** 380–385.

Wightman, B., Ha, I., and Ruvkun, G. (1993). Posttranscriptional regulation of the hetero-chronic gene *lin-14* by *lin-4* mediates temporal pattern formation in *C. elegans. Cell* **75,** 855–862.

Wulczyn, F. G., Smirnova, L., Rybak, A., Brandt, C., Kwidzinski, E., Ninnemann, O., Strehle, M., Seiler, A., Schumacher, S., and Nitsch, R. (2007). Post-transcriptional regulation of the *let-7* microRNA during neural cell specification. *FASEB J.* **21,** 415–426.

Yang, Z., Ebright, Y. W., Yu, B., and Chen, X. (2006). HEN1 recognizes 21–24 nt small RNA duplexes and deposits a methyl group onto the 2′ OH of the 3′ terminal nucleotide. *Nucleic Acids Res.* **34,** 667–675.

Yu, B., Yang, Z., Li, J., Minakhina, S., Yang, M., Padgett, R. W., Steward, R., and Chen, X. (2005). Methylation as a crucial step in plant microRNA biogenesis. *Science* **307,** 932–935.

Zhao, Y., Samal, E., and Srivastava, D. (2005). Serum response factor regulates a muscle-specific microRNA that targets Hand2 during cardiogenesis. *Nature* **436,** 214–220.

CHAPTER SEVEN

Cloning and Detecting Signature MicroRNAs from Mammalian Cells

Guihua Sun,*,† Haitang Li,† and John J. Rossi*,†

Contents

Abstract

MicroRNAs (miRNAs) are about 19- to 24-nucleotides long noncoding regulatory small RNAs that could silence target gene expression through base pairing to the complementary sequences in the 3′ untranslated region (3′UTR) of targeted genes. They are evolutionarily conserved and play an important regulatory role in embryogenesis, cell differentiation, and proliferation. They are also involved in pathogenesis and progression of some human diseases. There are about 1000 human miRNAs predicted today, and it is estimated that they could target about 30% of all human transcripts. Profiling the miRNAs that are expressed in the experimental cells became an important issue as different cells express different signature miRNAs or express the same miRNAs at different level. Small RNA cloning is a reliable way to characterize those tissue- or cell-specific signature miRNAs. This chapter describes a relatively nonlaborious polyadenylation-mediated

* Graduate School of Biological Sciences, Beckman Research Institute of the City of Hope, Duarte, California
† Division of Molecular Biology, Beckman Research Institute of the City of Hope, Duarte, California

Methods in Enzymology, Volume 427

ISSN 0076-6879, DOI: 10.1016/S0076-6879(07)27007-7

complementary DNA (cDNA) cloning method that will identify most of the small RNAs expressed in the cells of interest. This procedure can also be used to verify bioinformatic predictions of miRNAs/small interfering RNAs (siRNAs) as well as to identify new miRNAs/siRNAs.

1. Introduction

Small RNAs are gaining the attention for their regulatory role in development, viral pathogenesis, and progression of some human diseases. Among them, the miRNA family is the most extensively studied.

MiRNAs can derive from intronic or exonic RNA polymerase II transcripts with a cap and poly(A) tail-termed primary (pri)-miRNAs (Cai *et al.*, 2004; Lee *et al.*, 2002, 2003). Pri-miRNAs are further processed to precursor (pre)-miRNAs by the endoribonuclease (RNase) III Drosha that partners with the RNA-binding protein DGCR (DiGeorge Critical Region) 8 in the nucleus (Han *et al.*, 2004; Lee *et al.*, 2003). Pre-miRNAs will be exported to cytoplasm by expotin-5 (Yi *et al.*, 2003) and become mature miRNAs following cleavage by the RNase III enzyme Dicer, which partners with the RNA binding protein TRBP (TAR RNA binding protein) (Chendrimada *et al.*, 2005; Haase *et al.*, 2005). The mature miRNAs will be used to guide Argonaute family members in the RNA induced silence complex (RISC) in identifying the target sequences within the 3′UTRs of targeted genes for translational inhibition (Bartel, 2004). It may also repress gene expression by sequestering targeted mRNAs to mRNA processing bodies (P-bodies) (Liu *et al.*, 2005) for mRNA storage or decapping, deadenylation, and degradation.

In contrast to most plant miRNAs that are nearly perfectly complementary to their targeted sequences, most of the animal miRNAs make imperfect complementary base pairing to their targeted sequences. However, complete complementarity of seven nucleotides at the 5′ end of miRNAs positioned at nucleotides 2 to 8, the so called "seed sequence," to its targeted sequences has been shown to be crucial for an miRNA function. For those binding sites with imperfect seed sequences, a strong 3′ base pairing could compensate for mediating more potent miRNA function. Based on current miRNA targeting theories, it is estimated that, on average, an miRNA can target 200 different transcripts (Lewis *et al.*, 2005). There are about 472 cloned or computer-predicted human miRNAs (http://microrna.sanger.ac.uk/cgi-bin/sequences/mirna_summarypl?org=hsa) in the current human miRNA database, although, theoretically, it is estimated that there are about 1000 human miRNAs. This corresponds to approximately 2 to 3% of the human genome. So, about 20 to 25% of human transcripts can be targeted by miRNAs (Lewis *et al.*, 2005).

Because the first miRNA was discovered in *Caenorhabditis elegant* and was found to play an essential role in the timing of worm's development (Lee *et al.*, 1993; Wightman *et al.*, 1993), it is now a wildly accepted concept that miRNAs are important regulatory elements in development, apoptosis, disease generation, and progression. They are small, but could perform functions equivalent to major transcription factors. Generally, they silence transcripts by guiding RISC binding to 3′UTR of targeted gene, disrupting translation initiation, resulting in targeted translational inhibition (Doi et al., 2003; Humphreys et al., 2005; Pillai et al., 2005). miRNAs may also function to direct targeted transcripts to P-bodies for temporary sequestering from translation, and subsequently, these transcripts could be reshuttled back to the cytoplasm to participate again as translation templates (Bhattacharyya et al., 2006). Alternatively, they could sequester transcripts to P-bodies for decapping, deadenylation, and degradation (Chu and Rana, 2006; Liu et al., 2005; Rehwinkel et al., 2005). miRNAs and their associated RISC proteins can fine-tune gene expression in addition to that which is imparted by transcription factors. Transcription factors themselves are major targets of miRNAs. By targeting transcription factors, miRNAs could form gene expression regulatory circuits with transcription factors to regulate target gene expression.

There are four major ways to profile miRNA expression in cells. Northern hybridization, miRNA cloning, miRNA microarray, and quantitative RT-PCR. Multiple techniques usually are engaged to verify the results. The small size of miRNAs is a challenge for miRNA detection. Northern hybridization remains as the major technique to detect miRNA expression, but often requires large amounts of total RNA, which may be impractical with some cell types. For low-abundance small RNAs, they may not be detected even with more than 30 μg of total RNA using traditional DNA oligonucleotide probes, although the use of somewhat expensive backbone modified probes, such as locked nucleic acids (LNA) (Valoczi et al., 2004), may give more sensitivity than conventional DNA oligos.

During the last 2 years, several large-scale profiling techniques including miRNA microarrays (Barad et al., 2004; Castoldi et al., 2006; Liang et al., 2005; Liu et al., 2004; Nelson et al., 2004, 2006; Thomson et al., 2004; Yeung et al., 2005) and quantitative reverse-transcription mediated polymerase chain reaction (qRT-PCR) have been reported for miRNA analyses (Chen et al., 2005). miRNA microarrays and qRT-PCR are much better large-scale profiling techniques than Northern blotting and need relatively very small amounts of RNA samples. Their limitation is that they can only detect known miRNAs, and some probes will only detect pre-miRNAs, which do not always correspond to the expression level of mature miRNAs. Applying the same hybridization conditions for all of the miRNAs may also be problematic.

Cloning is a reliable technique to identify novel miRNAs and to verify in silico–predicted miRNAs. Additionally cloning can be used to verify the ends of the mature miRNA sequences (Ambros and Lee, 2004; Fu et al., 2005; Lagos-Quintana et al., 2002; Lau et al., 2001; Lee and Ambros, 2001; Takada et al., 2006). A combination of miRNA microarray and cloning techniques have been reported to effectively identify many novel miRNAs (Berezikov et al., 2006). A massively parallel, high-throughput sequencing technique developed by 454

Life Sciences Corporation allows large-scale sequence determination of total small RNA populations (Aravin *et al.*, 2006; Girard *et al.*, 2006; Lau *et al.*, 2006).

Most of the miRNA cloning techniques that have been developed in the past several years are time consuming, labor intensive, and often fail in the small RNA fractionation step. Small RNA fractionation is a critical step in miRNA cloning. Usually, RNA is electrophoresed in denaturing polyacrylamide gel (PAGE). Radio-labeled RNA oligonucleotides are used to determine the electrophoretic mobility of the small RNAs. The small RNAs ranging in size between 19 to 24 nucleotides (nt) in length are excised and eluted from the gels. They are then joined to 5$'$ and 3$'$ adaptors by T4 RNA ligase followed by first-strand cDNA reverse transcription and subsequent PCR amplification. A report has described a polyadenylation, small RNA cloning method that eliminated the gel purification step (Fu *et al.*, 2005). This approach represents a relatively simple miRNA cloning technique.

The miRNA cloning protocol in this chapter was based upon a combination of methods to isolate small RNAs by flashPAGE, developed by Ambion Inc., as well as protocols from the Dr. Bartel laboratory (Lau *et al.*, 2001), the Ambros laboratory (Ambros and Lee, 2004; Lee and Ambros, 2001) and polyadenylation cloning procedure reported by Fu *et al.* (2005) (Fig. 7.1). This protocol eliminates the gel purification step, which can result in poor miRNA cloning results. The removal of the 5$'$ phosphate group from the small RNAs in some protocols can result in RNA degradation, base damage, and loss of terminal bases. The only gel purification step in this protocol is for amplified cDNA fragment purification. Due to the addition of a 5$'$ adaptor and polyA tail, the cDNAs will range between 100 and 110 base pairs, which can be stained with Ethidium bromide, allowing identification and relatively straightforward purification.

Cloning of small RNAs can also be problematic in that miRNAs cannot be readily distinguished from other small RNA and DNA fragments. Usually the cloning results should be verified by Northern analyses or miRNA array analyses. For previously uncharacterized miRNAs that have not been previously identified by other methods, a low-energy folding structure generated by an RNA folding program, such as mFOLD (Zuker, 2003) or Vienna RNAfold (Hofacker 2003), can be used to predicate precursor hairpin structures from the bioinformatics analyses. A flow diagram for cloning is presented in Fig. 7.1.

2. miRNA Cloning

2.1. Total RNA isolation

Using RNA STAT-60 (Tel-Test Inc.) to isolate total RNA for small RNA fractionation is highly recommended. All good RNA working habits should be followed.

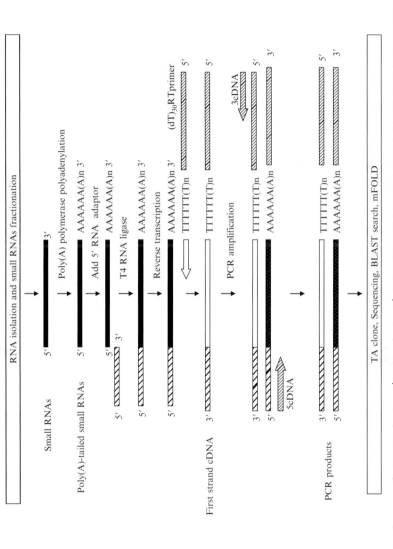

Figure 7.1 Flow diagram of miRNA cloning. 5′ RNA adaptor: 5′-CGA CUG GAG CAC GAG GAC CAC GAG GAC ACU GAC AUG GAC UGA AGG AGU AGA AA-3′ (dT)$_{30}$RTprimer: ATT CTA GAG GCC GAG GCG GCC GAC ATG-d(T)$_{30}$ (A, G, or C) (A, G, C, or T) 5cDNA: 5′-GGA CAC TGA CAT GGA CTG AAG GAG TA-3′ 3cDNA: 5′-ATT CTA GAG GCC GAG GCG GCC GAC ATG T-3′.

1. Remove medium from plate for adhesion cells or spin down cells, then remove medium for suspension cells. It is not necessary to wash the cells with PBS, which may result in RNA degradation.
2. Add proper amount (1 ml for 1×10^7 cells) of RNA STAT-60 to the plate or the cell pellet. Vigorously pipette up and down until the viscosity disappears. You may need to add more STAT-60 to achieve this result. This step is critical to get high-quality RNA.
3. Add 0.2 ml chloroform to every 1 ml STAT-60 lysate, let it sit on the bench for 3 to 5 min, spin at 12,000g, 4° for 15 min.
4. Transfer the aqueous phase (about 0.6 ml) to a new tube, add 0.5 ml isopropanol, mix, and let it sit on bench at room temperature for 3 to 5 min. Spin at 4°, 10,000g for 15 min.
5. Carefully remove all the liquid, add 1-ml 75% ethyl alcohol (EtOH) that was made with diethylpyrocarbonate (DEPC)-treated water (DEPC-H_2O, Ambion Inc.) to wash the pellet. Spin at 4°, 7500g for 15 min.
6. Carefully remove all the liquid and air-dry the pellet. (It is a good idea to always dry RNA sample in the cell culture hood whenever possible).
7. Dissolve the pellet in the proper amount of DEPC-H_2O. It is recommended to dissolve total RNA at the concentration between 1 and 2 $\mu g/\mu l$ to facilitate the purification of small RNAs by flashPAGE gel. Usually you can get about 100 μg total RNA from 1×10^7 cells, so use 50 μl DEPC-H_2O to dissolve total RNAs from one confluent 10-cm plate. Measure the concentration and check the quality by measuring the UV absorbance at 260 nm.
8. Total RNA quality could be checked by running on a denatured agarose gel followed by staining with ethidium bromide (EtBr, 10 mg/ml). It may be necessary to remove contaminated DNA by treating the total RNA with RNase-free DNase I.

2.2. Fractionize small RNA with the flashPAGE™ fractionator system

The flashPAGE Fractionator System from Ambion provides a convenient, efficient alternative to the laborious and time-consuming PAGE and subsequent small RNAs elution from crushed gel that damage bases in RNA or degrade RNA samples.

1. Clean the upper chamber, lower chamber, and the electronode with RNaseZap® RNase decontamination solution (Ambion, Inc.). Rinse thoroughly with DEPC-H_2O.
2. Load 250-μl lower chamber buffer from flashPAGE buffer kit (Type A, Ambion, Inc.) to the lower chamber, mount a flashPAGE gel, then load upper chamber with 250 μl gel running buffer from flashPAGE buffer kit.
3. Prepare 10 to 100 μg total RNA in 50 μl DEPC-H_2O. (The capacity for flashPAGE is 100 μg, multiple gels should be used for over 100 μg

total RNA or the RNA concentration will be too low to limit the total volume to 50 μl. The same lower running buffer could be used if more than one gel is needed to keep the precipitation volume small.) Add 50 μl A40 tracking dye from flashPAGE buffer kit, which will run at the same speed as 40-nt long single-strand RNAs. Heat the sample at 95° for 5 min, chill the sample on ice, then pipette the whole sample into the upper chamber buffer.

4. Run the fractionator at 75 V for about 13 min. Stop it when the blue dye just starts coming out of the bottom surface of the flashPAGE gel.
5. Collect the entire lower chamber buffer, which now contains the small RNAs shorter than 40 nt. Add 250 μl mixture of phenol/chloroform/isoamyl alcohol (25:24:1, pH 6.6, Ambion), mix and spin at 12,000g for 5 min. Collect the aqueous phase, and add 25 μl (0.1 vol), pH 5.2, 3 M sodium acetate, 625 μl (2.5 vol) 100% EtOH, 1 μl glycogen (20 μg/μl, Roche). Precipitate at least 2 h at −20° (overnight is strongly suggested).

2.3. Polyadenylation of small RNAs

Polyadenylation of small RNAs serves four purposes. First, it will stabilize the small RNA. Second, it serves as a priming site for synthesis of first-strand cDNA using a primer with poly(dT) on its 3′-end portion. Third, it will eliminate the step to remove the 5′ phosphate from small RNA, the ligation step of 3′ adaptor, and subsequent step to remove excessive 3′ adaptor from the ligation products by gel purification. Fourth, it will make the first-strand cDNA longer (over 100 bases long), which will facilitate the next step, purification of amplified cDNA products from PAGE gel.

1. Spin at 12,000g at 4° for 15 min to pellet precipitate small RNA.
2. Carefully remove all the liquid, avoiding touching the pellet. Add 1 ml 75% EtOH to the tube to wash the pellet. Spin at 10,000g at 4° for 15 min.
3. Carefully remove all the liquid, avoiding touching the pellet. Air-dry the pellet at room temperature. Dissolve the pellet in 40 μl DEPC-H$_2$O (20 μl could be saved at −80° and used for the polyadenylation reaction).
4. Polyadenylation was performed with an A-Plus^{TM} Poly(A) Polymerase Tailing Kit (EPICENTRE® Biotechnologies). The following polyadenylation protocol is used for small RNAs fractionized from 50 to 100 μg total RNA (about 5 to 10 ng small RNAs): add 5 μl 10× polyadenylation buffer, 5 μl 10 mM ATP, 2 μl RNase-out, 1 μl poly(A) polymerase to the 40 μl small RNAs (add 20 μl DEPC-H$_2$O if you only used half of the purified small RNAs, RNase-out is from Invitrogen) to make it final reaction volume 50 μl. Incubate at 37° for 30 min.
5. Add 50 μl phenol/chloroform/isoamyl alcohol (25:24:1, pH 6.6) to the reaction mixture. Mix, then spin at 12,000g at 4° for 5 min.
6. Collect the aqueous phase, add 5 μl (0.1 vol), pH 5.2, 3 M sodium acetate, add 125 μl (2.5 vol) 100% EtOH, and add 1 μl glycogen

(20 μg/μl, Roche). Precipitate polyadenylated small RNA for at least 2 h at $-20°$. Precipitation overnight is preferred.

2.4. Ligation of 5′ adaptor to polyadenylated small RNA

The 5′ adaptor will be used as the priming site for amplification of first-strand cDNA. It also could be designed with restriction sites to concatemerize amplification products for cloning. 5′ adaptor is RNA and the sequence is 5′-CGA CUG GAG CAC GAG GAC ACU GAC AUG GAC UGA AGG AGU AGA AA-3′.

1. Spin precipitated polyadenylated small RNA from step 2.3.6 at 12,000g for 15 min.
2. Carefully remove the supernatant and add 1 ml 75% EtOH to wash the pellet. Spin at 10,000g at 4° for 15 min.
3. Carefully remove all the liquid, avoiding touching the pellet. Air-dry the pellet at room temperature.
4. Dissolve the dried pellet in 10 μl DEPC-H$_2$O (5 μl could be taken out and saved at $-80°$; use the remaining 5 μl for adding 5′ adaptor, add 5 μL DEPC-H$_2$O to make it 10 μl).
5. Add 2 μl 200-μM 5′ adaptor, 4 μl 5 × RNA ligation buffer, 2 μl RNA ligase (NEB), 1 μl RNase-Out (Invitrogen). Incubate at room temperature for 3 h.
6. Add 20-μl phenol/chloroform/isoamyl alcohol (25:24:1, pH 6.6) to the reaction mixture. Mix, then spin at 12,000g at 4° for 15 min.
7. Collect the supernatant, add 2 μl (0.1 vol) pH 5.2, 3 M sodium acetate, add 50 μl (2.5 vol) ethyl alcohol, and add 1 μl glycogen (20 μg/μl). Precipitate 5′ adaptor-small RNA-poly(A)n at least 2 h at $-20°$ (overnight is preferred).

2.5. Reverse transcription

Once you get the 5′ adaptor-small RNA-poly(A)n, a primer with (dT)$_{30}$ could be used to bind poly(A) tail and prime the reverse transcription (RT). Primer binding site for cDNA amplification was added following the (dT)$_{30}$. This primer is called (dT)$_{30}$RTprimer: ATT CTA GAG GCC GAG GCG GCC GAC ATG-d(T)$_{30}$VN. V is anyone of A, G, or C, and N is anyone of A, G, C, or T (V and N will bind to the two nt in the 3′ end of small RNA).

1. Spin precipitated polyadenylated small RNA with 5′ adaptor from step 2.4.7 at 12,000g for 15 min at 4°.
2. Carefully remove the supernatant, add 1 ml 75% EtOH to wash the pellet. Spin at 10,000g for 15 min at 4°.
3. Carefully remove all the liquid, avoiding touching the pellet. Air-dry the pellet at room temperature.
4. Dissolve the dried pellet in 10 μl DEPC-H$_2$O (5 μl could be take out and saved at $-80°$; use the remaining 5 μl for RT reaction).

5. Add 1 μl 10-mM dNTPs and 1 μl (dT)$_{30}$RTprimer (50 μM) to the earlier 10 μl 5' adaptor-small RNA-poly(A)n (add 5-μl DEPC-H$_2$O if you used half of the 5' adaptor-small RNA-poly(A)n). Incubate at 65° for 5 min and then chill on ice at least 1 min.

6. Add the following from first-strand cDNA synthesis kit (Invitrogen) to the previous tube: 2 μl 10× RT buffer, 4 μl 25 mM Mg^{++}, 2 μl 0.1 M DTT, 1 μl RNase-Out, 1 μl Superscript III. Incubate the reaction tube at 50° for 50 min.

7. Incubate the reaction mixture at 85° for 5 min to inactivate the reaction.

8. Add 1 μl RNase H to the RT reaction tube and incubate at 37° for 20 min to degrade RNA in the reaction mixture.

9. One or 3 μl of the RT products could be run on an 8-M urea, 15% PAGE gel (Bio-Rad mini-gel system) and stained with SYBR gold nucleic acid gel stain (Invitrogen) to check the RT products.

2.6. Amplification of first-strand cDNA by PCR

A primer set in which one base pairs with 5' adaptor and one base pairs with 3' poly(A)n is used to amplify the small RNA cDNA. 5cDNA (5'-GGA CAC TGA CAT GGA CTG AAG GAG TA-3') and 3cDNA (5'-ATT CTA GAG GCC GAG GCG GCC GAC ATG T-3') were used as the primer set. (We also used a set of primers with Ban I sites to concatamerize the PCR products for cloning to reduce the sequencing cost. The primer sets are 5Ban-cDNA: 5'-ATC GTA GGC ACC GGA CAC TGA CAT GGA CTG AAG GAG TA-3' and 3Ban-cDNA: 5'-ATT GAT GGT GCC ATT CTA GAG GCC GAG GCG GCC GAC ATG T-3').

1. Set up the reaction by adding 5 μl of the first-strand cDNA from step 2.5.8, 1 μl 5cDNA (50 μM), 3cDNA (50 μM), and then add 45 μl Platinum PCR mix (Invitrogen).

2. Amplify with the following program: one cycle at 94° for 2 min to denature the template; 30 cycles at 94° for 30 sec, 50° for 30 sec, 72° for 30 sec; extend the reaction for one cycle at 72° for 5 min.

2.7. PAGE gel purification of amplified cDNA

Eight-M urea, 12% PAGE gel was used to separate PCR band. The PCR products that contain small RNA will be 26 nt (5cDNA) + 19 to 24 nt long small RNAs + poly(A)30 + 28 nt (3cDNA) long. So it is between 103 and 108 bp (Fig. 7.2). If the Ban I set of primer is used, it will be around 130-bp long.

1. Prepare a 12% acrylamide, 8-M urea gel with Bio-Rad minigel system. Add 10 μl 5× Orange G DNA loading dye to the amplified cDNA sample (Orange G will run out of the gel, so it will not interfere with

viewing the bands and subsequent cutting of the bands). Load the sample to six lanes with 10 μ each. Run 10-bp (Invitrogen) and 100-bp (New England Biolab) DNA molecular weight ladders (in Xylene cyanol/Bromophenol blue loading dye) on the side. The 100-bp DNA ladder will help locating the 100-bp band in the 10-bp DNA ladder if the gel is run too long and the smaller bands from 10-bp DNA ladder runs out of the gel.

2. Run gel until the Xylene cyanol reaching two-thirds length of the gel or after Bromophenol blue is just running out of the bottom of the gel. Stain gel with EtBr or SYBR gold nucleic acid gel stain.

3. Locate the band between 100 bp and 110 bp (if the Ban I primer set was used the band will be around 130 bp), cut the band and smash it to very small pieces. (We find it is convenient to cut them on a clean X-ray film with a clean razor blade.)

4. Transfer the smashed gel to a 1.5-ml tube and add 1 ml elution buffer (0.5 M NH$_4$OAc, 10 mM MgCl$_2$, 1 mM EDTA). Elute at room temperature or 37° overnight on a rotating device.

5. Divide the 1-ml eluate into two tubes with 0.5 ml each, add 0.5 ml phenol/chloroform/isoamyl alcohol (25:24:1, pH 7.9, Ambion) to each tube, and spin the tube at 12,000g for 5 min.

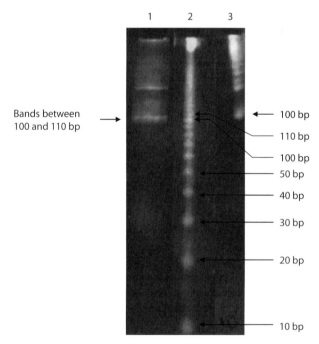

Figure 7.2 PAGE gel of amplified cDNA. Lane 1: Amplified sample. Lane 2: 10-bp DNA ladder. Lane 3: 100-bp DNA ladder.

6. Transfer the aqueous phase to a new tube, add 50 μL (0.1 vol), pH 7.0, 3 M NaOAc and 1 ml (2 vol) 100% EtOH. Precipitate amplified cDNA in the aqueous phase over 30 min at $-20°$.

2.8. TOPO-TA cloning of the PCR products

It is important to use this high-efficiency PCR products cloning vector system to get a higher titer-cloned small RNA library. One-Shot TOP10 chemically competent *Escherichia coli* should be used for this purpose too. It will allow for blue/white screen-positive clones, and the blue/white selection is also a convenient indicator of the quality of the library. A good library will have thousands of colonies, and about 90% of them will be white.

1. Spin the precipitation from step 2.7.6 at 12,000 g for 15 min.
2. Carefully remove the supernatant, add 1 ml 70% EtOH to wash the pellet. Spin at 10,000g for 15 min.
3. Carefully remove all the liquid, avoiding touching the pellet. Air-dry the pellet at room temperature.
4. Dissolve the dried pallet in 50 μl TE (10 mM Tris-HCl, 1 mM EDTA) buffer or dH$_2$O.
5. Pool these eluates into one tube and purify it with PCR products purification kit (QIAGEN). Elute the column with 30 μl dH$_2$O. Run 3 μl of the eluate on a 1.5% agarose gel to check the purified products as one discrete band on the gel. Measure the concentration of the DNA in the eluate by reading absorbance at O.D. 260.
6. Based on the amount of DNA present in the gel, set up the following TA ligation: 1 μl TA vector (TOPO-TA cloning kit, Invitrogen), 1 μl 10$\times$ ligation buffer, 1 to 5 μl PCR products, add water to make the final volume 10 μl, and add 1 μl Ligase. Ligate at 14° overnight.
7. Combine one tube of One-Shot TOP10 competent cells (Invitrogen) with 2 μl of the previous ligation products and use X-gal/IPTG to screen white colonies as colonies with insert.
8. Pick up 10 white colonies to make plasmid DNA and sequence them with M13 primers. This will ensure there are inserts in the white colonies before large numbers of colonies are going to be sequenced. Sequence 1000 colonies. You should be able to get about 300 to 400 miRNA colonies. About 100 clones could be unique miRNA clones and will give you a pretty good idea about the miRNAs predominantly expressed in your favorite cells.

3. MIRNA IDENTIFICATION

The positive clones with insert should be easy to spot. It will be in front of a stretch of As, after the 5' adaptor, or after a stretch of Ts, in front of reversed 5' adaptor. You may want to use software, such as Sequencher

(Gene codes corporation), to trim the sequences from TA-clone vector and the sequences from $5'$ adaptor and poly(A) or poly(T). The procedure described in this chapter is intended for individuals with little bioinformatics background. A Perl or other software script to retract the sequences will greatly reduce your workload. It is a good idea to seek help from individuals with a bioinformatics background during these steps.

1. Open the web page: http://microrna.sanger.ac.uk/sequences/search.shtml. Paste your sequence into the "search by sequence" box, click the "search miRNAs" button. This will identify all annotated miRNAs in the miRBase.

2. If it does not exist in miRBase, they are probably tRNAs, rRNAs, snoRNAs, piRNAs, fragments of degraded mRNAs, or unidentified miRNAs. Open NCBI Blast search web page at http://www.ncbi.nlm. nih.gov/BLAST/. Click on "Search for short, nearly exact matches" button. Paste the sequence into the search box. If they do not belong to known tRNAs, rRNAs, snoRNAs, piRNAs, or mRNAs, they are probably unidentified miRNAs.

3. Open the ensemble genome browser at http://www.ensembl.org/index. html, click "Run a BLAST Search" button under "Use Ensemble to..." on the left side of the page. Paste the sequence into the query sequence box. Use the default setting, except select your sample species (the default is human) and change the search sensitivity from "Near-exact matches" to "Exact matches." Click the red "Run?" and wait for the program to finish parsing the result and the red "View?" to appear. Click the red "View?" to open the results window and click on the match square on the chromosome. In the pop-up menu, click "Genomic Sequence...." The newly opened web page will show a stretch of genome sequence flanking 300 bases on each side of your sequence. First change the $5'$ flanking sequence to 65 and $3'$ flanking sequence to 15, then click the update button. Copy the updated sequence. Open mFold web page at http:// www.bioinfo.rpi.edu/applications/mfold/rna/form1.cgi. Paste the sequence into the sequence enter box using the "all default" setting. Click "Fold RNA" button. Save the predicted hairpin structure. Next, change the $5'$ flanking sequence to 15 and $3'$ flanking sequence to 65 and perform the same steps as before to get the folding structure. Follow the several rules used to evaluate the predicated structure previously published by Ambros and Lee (2004; Lee and Ambros, 2001, 2004). Novel miRNAs can be registered at http://microrna.sanger.ac.uk/registry/.

4. Northern Hybridization to Verify *In Vivo* Expression of miRNA

Northern blot usually used to verify the cloned miRNAs do present *in vivo*. But, Northern blot will not be able to detect all of them, especially those expressed at very low levels. If a DNA oligonucleotide probe does not work, a LNA oligonucleotide probe that will give more sensitivity and higher specificity can be used for detection.

The Northern blot protocol for detecting siRNA from Li and Rossi (2005) can be used to detect miRNA. We routinely use 10 or 15% PAGE, 8-M urea denatured gel for miRNA Northern blot (Fig. 7.3). A 10% gel gives a good separation and is more efficient for transfer to membrane, but a 15% gel can give a better resolution and better detection.

1. Mix 20 μg total RNA with equal volume loading buffer (95% deionized formamide, 0.025% bromophenol blue, 0.025% xylene cyanol, 0.5 mM EDTA, 0.025% SDS). Heat the sample at 90° for 5 min and then chill on ice.

2. Make a gel about 10 cm long, load the sample, load γ^{32}P-labeled decade RNA maker (Ambion). Run gel at 250 V (25 V/cm) for 2 to 3 h or until bromophenol blue dye migrates 1 to 2 inches closer to the bottom of the gel.

3. Follow the gel transfer procedure of GENIE® Electrophoretic transfer's protocol from IDEA scientific company. Use 0.5× TBE as transfer (1× TBE is OK). Use Hybond-N+ (Amersham Pharmacia biotech, positively charged; not Hybond-N-membrane, negatively charged) membrane. Transfer at 0.75 mA for 1 h. UV cross-link the membrane twice with auto–cross-linking setting (Stratalinker® 2400 UV Crosslinker, Stratagene).

4. Prehybridize the membrane in buffer containing 6× SSPE (20× SSPE stock: 3 M NaCl, 0.2 M NaH$_2$PO$_4$, 0.02 M EDTA, pH 7.4), 5×

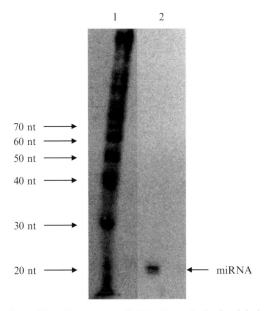

Figure 7.3 Northern blot detecting miRNA. Lane 1: Radio-labeled decade RNA maker. Lane 2: RNA sample with detected miRNA.

Denharts, 0.5% SDS, 50% formamide, and carrier DNA (10 $\mu g/\mu l$, 250 to 350 μl for 15-ml prehybridization buffer) at 37° for 2 h. Alternatively, PerfectHyb^{TM} Plus Hybridization Buffer from Sigma is an excellent buffer for prehybridization and hybridization. It is much more sensitive than conventional buffers.

5. Label 1 μl (10 pmol) 10-μM DNA oligonucleotide probe complementary to the antisense sequence of the miRNA with γ^{32}P-ATP in 10-μl reaction volume. Dilute to 50 μl and pass through a G–25 column (Amersham). Denature 15 μl of the probe by heating at 95° for 5 minutes. Add to the prehybridization buffer and hybridize overnight at 37°.

6. Wash the membrane with 6× SSPE and 0.1% SDS at 37° for 10 min then wash with 2 × SSC (20 × SSC stock: 3 M NaCl, 0.3 M NaCitrate, pH 7.0) and 0.1% SDS twice at 37° for 10 min each.

7. Drain and wrap in cellophane wrap and expose to X-ray film, keeping the blot moist. Blots with higher background could be washed again.

8. Blots could be restriped with 0.1% SDS, 0.1× SSC at 85 to 90° for 10 to 20 min to hybridize with other probes.

ACKNOWLEDGMENTS

We are very grateful to Dr. Jianfei Ji, Dr. Ming-Jie Li, Dr. Mohammed Amarzguioui, Dr. Harris Hoffer, and Dr. Lars Aagaard in the Rossi lab for their technical expertise. This work was supported by NIH grants from NIAID and HLB to JJR.

REFERENCES

Ambros, V., and Lee, R. C. (2004). "Identification of microRNAs and other tiny noncoding RNAs by cDNA cloning." *Methods Mol. Biol.* **265,** 131–158.

Aravin, A., Gaidatzis, D., Pfeffer, S., Lagos-Quintana, M., Landgraf, P., Iovino, N., Morris, P., Brownstein, M. J., Kuramochi-Miyagawa, S., Nakano, T., Chien, M., Russo, J. J., *et al.* (2006). "A novel class of small RNAs bind to MILI protein in mouse testes." *Nature* **442**(7099), 203–207.

Barad, O., Meiri, E., Avniel, A., Aharonov, R., Barzilai, A., Bentwich, I., Einav, U., Gilad, S., Hurban, P., Karov, Y., Lobenhofer, E. K., Sharon, E., *et al.* (2004). "MicroRNA expression detected by oligonucleotide microarrays: System establishment and expression profiling in human tissues." *Genome Res.* **14**(12), 2486–2494.

Bartel, D. P. (2004). "MicroRNAs: Genomics, biogenesis, mechanism, and function." *Cell* **116**(2), 281–297.

Berezikov, E., van Tetering, G., Verheul, M., van de Belt, J., van Laake, L., Vos, J., Verloop, R., van de Wetering, M., Guryev, V., Takada, S., van Zonneveld, A. J., Mano, H., *et al.* (2006). "Many novel mammalian microRNA candidates identified by extensive cloning and RAKE analysis." *Genome Res.* **16**(10), 1289–1298.

Bhattacharyya, S. N., Habermacher, R., Martine, U., Closs, E. T., and Filipowicz, W. (2006). "Relief of microRNA-mediated translational repression in human cells subjected to stress." *Cell* **125**(6), 1111–1124.

Cai, X., Hagedorn, C. H., and Cullen, B. R. (2004). "Human microRNAs are processed from capped, polyadenylated transcripts that can also function as mRNAs." *RNA* **10**(12), 1957–1966.

Castoldi, M., Schmidt, S., Benes, V., Noerholm, M., Kulozik, A. E., Hentze, M. W., and Muckenthaler, M. U. (2006). "A sensitive array for microRNA expression profiling (miChip) based on locked nucleic acids (LNA)." *RNA* **12**(5), 913–920.

Chen, C., Ridzon, D. A., Broomer, A. J., Zhou, Z., Lee, D. H., Nguyen, J. T., Barbisin, M., Xu, N. L., Mahuvakar, V. R., Andersen, M. R., Lao, K. Q., Livak, K. J., et al. (2005). "Real-time quantification of microRNAs by stem-loop RT-PCR." *Nucleic Acids Res.* **33**(20), e179.

Chendrimada, T. P., Gregory, R. I., Kumaraswamy, E., Norman, J., Cooch, N., Nishikura, K., and Shiekhattar, R. (2005). "TRBP recruits the Dicer complex to Ago2 for microRNA processing and gene silencing." *Nature* **436**(7051), 740–744.

Chu, C. Y., and Rana, T. M. (2006). "Translation Repression in Human Cells by Micro-RNA-Induced Gene Silencing Requires RCK/p54." *PLoS Biol.* **4**(7), e210.

Doi, N., Zenno, S., Ueda, R., Ohki-Hamazaki, H., Ui-Tei, K., and Saigo, K. (2003). "Short-interfering-RNA-mediated gene silencing in mammalian cells requires Dicer and eIF2C translation initiation factors." *Curr. Biol.* **13**(1), 41–46.

Fu, H., Tie, Y., Xu, C., Zhang, Z., Zhu, J., Shi, Y., Jiang, H., Sun, Z., and Zheng, X. (2005). "Identification of human fetal liver miRNAs by a novel method." *FEBS Lett.* **579**(17), 3849–3854.

Girard, A., Sachidanandam, R., Hannon, G. J., and Carmell, M. A. (2006). "A germline-specific class of small RNAs binds mammalian Piwi proteins." *Nature* **442**(7099), 199–202.

Haase, A. D., Jaskiewicz, L., Zhang, H., Laine, S., Sack, R., Gatignol, A., and Filipowicz, W. (2005). "TRBP, a regulator of cellular PKR and HIV-1 virus expression, interacts with Dicer and functions in RNA silencing." *EMBO Rep.* **6**(10), 961–967.

Han, J., Lee, Y., Yeom, K. H., Kim, Y. K., Jin, H., and Kim, V. N. (2004). "The Drosha-DGCR8 complex in primary microRNA processing." *Genes Dev.* **18**(24), 3016–3027.

Hofacker, I. L. (2003). "Vienna RNA secondary structure server." *Nucleic Acids Res.* **31**(13), 3429–3431.

Humphreys, D. T., Westman, B. J., Martin, D. J., and Preiss, T. (2005). "MicroRNAs control translation initiation by inhibiting eukaryotic initiation factor 4E/cap and poly(A) tail function." *Proc. Natl. Acad. Sci. USA* **102**(47), 16961–16966.

Lagos-Quintana, M., Rauhut, R., Yalcin, A., Meyer, J., Lendeckel, W., and Tuschl, T. (2002). "Identification of tissue-specific microRNAs from mouse." *Curr. Biol.* **12**(9), 735–739.

Lau, N. C., Lim, L. P., Weinstein, E. G., and Bartel, D. P. (2001). "An abundant class of tiny RNAs with probable regulatory roles in *Caenorhabditis elegans.*" *Science* **294**(5543), 858–862.

Lau, N. C., Seto, A. G., Kim, J., Kuramochi-Miyagawa, S., Nakano, T., Bartel, D. P., and Kingston, R. E. (2006). "Characterization of the piRNA complex from rat testes." *Science* **313**(5785), 363–367.

Lee, R. C., and Ambros, V. (2001). "An extensive class of small RNAs in *Caenorhabditis elegans.*" *Science* **294**(5543), 862–864.

Lee, R. C., Feinbaum, R. L., and Ambros, V. (1993). "The *C. elegans* heterochronic gene *lin-4* encodes small RNAs with antisense complementarity to *lin-14.*" *Cell* **75**(5), 843–854.

Lee, Y., Ahn, C., Han, J., Choi, H., Kim, J., Yim, J., Lee, J., Provost, P., Radmark, O., Kim, S., and Kim, V. N. (2003). "The nuclear RNase III Drosha initiates microRNA processing." *Nature* **425**(6956), 415–419.

Lee, Y., Jeon, K., Lee, J. T., Kim, S., and Kim, V. N. (2002). "MicroRNA maturation: Stepwise processing and subcellular localization." *EMBO J.* **21**(17), 4663–4670.

Lewis, B. P., Burge, C. B., and Bartel, D. P. (2005). "Conserved seed pairing, often flanked by adenosines, indicates that thousands of human genes are microRNA targets." *Cell* **120**(1), 15–20.

Li, M. J., and Rossi, J. J. (2005). "Lentiviral vector delivery of recombinant small interfering RNA expression cassettes." *Methods Enzymol* **392**, 218–226.

Liang, R. Q., Li, W., Li, Y., Tan, C. Y., Li, J. X., Jin, Y. X., and Ruan, K. C. (2005). "An oligonucleotide microarray for microRNA expression analysis based on labeling RNA with quantum dot and nanogold probe" *Nucleic Acids Res.* **33**, e17.

Liu, C. G., Calin, G. A., Meloon, B., Gamliel, N., Sevignani, C., Ferracin, M., Dumitru, C. D., Shimizu, M., Zupo, S., Dono, M., Alder, H., Bullrich, F., *et al.* (2004). "An oligonucleotide microchip for genome-wide microRNA profiling in human and mouse tissues." *Proc. Natl. Acad. Sci. USA* **101**(26), 9740–9744.

Liu, J., Rivas, F. V., Wohlschlegel, J., Yates, J. R., 3rd, Parker, R., and Hannon, G. J. (2005). "A role for the P-body component GW182 in microRNA function." *Nat. Cell. Biol.* **7**(12), 1261–1266.

Liu, J., Valencia-Sanchez, M. A., Hannon, G. A., and Parker, R. (2005). "MicroRNA-dependent localization of targeted mRNAs to mammalian P-bodies." *Nat. Cell. Biol.* **7**(7), 719–723.

Nelson, P. T., Baldwin, D. A., Kloosterman, W. P., Kauppinen, S., Plasterk, R. H., and Mourelatos, Z. (2006). "RAKE and LNA-ISH reveal microRNA expression and localization in archival human brain." *RNA* **12**(2), 187–191.

Nelson, P. T., Baldwin, D. A., Scearce, L. M., Oberholtzer, J. C., Tobias, J. W., and Mourolatos, Z. (2004). "Microarray-based, high-throughput gene expression profiling of microRNAs." *Nat. Methods* **1**(2), 155–161.

Pillai, R. S., Bhattacharyya, S. N., Artus, C. G., Zoller, T., Cougot, N., Basyuk, E., Bertrand, E., and Filipowicz, W. (2005). "Inhibition of translational initiation by *Let-7* MicroRNA in human cells." *Science* **309**(5740), 1573–1576.

Rehwinkel, J., Behm-Ansmant, I., Gatfield, D., and Izaurralde, E. (2005). "A crucial role for GW182 and the DCP1:DCP2 decapping complex in miRNA-mediated gene silencing." *RNA* **11**(11), 1640–1647.

Takada, S., Berezikov, E., Yamashita, Y., Lagos-Quintana, M., Kloosterman, W. P., Enomoto, M., Hatanaka, H., Fujiwara, S., Watanabe, H., Soda, M., Choi, Y. L., Plasterk, R. H., *et al.* (2006). "Mouse microRNA profiles determined with a new and sensitive cloning method." *Nucleic Acids Res.* **34**(17), e115.

Thomson, J. M., Parker, J., Perou, C. M., and Hammond, S. M. (2004). "A custom microarray platform for analysis of microRNA gene expression." *Nat. Methods* **1**(1), 47–53.

Valoczi, A., Hornyik, C., Varga, N., Burgyan, J., Kauppinen, S., and Havelda, Z. (2004). "Sensitive and specific detection of microRNAs by Northern blot analysis using LNA-modified oligonucleotide probes." *Nucleic Acids Res.* **32**(22), e175.

Wightman, B., Ha, I., and Ruvkun, G. (1993). "Posttranscriptional regulation of the heterochronic gene *lin-14* by *lin-4* mediates temporal pattern formation in *C. elegans.*" *Cell* **75**(5), 855–862.

Yeung, M. L., Bennasser, Y., Myers, T. G., Jiang, G., Benkirane, M., and Jeang, K. T. (2005). "Changes in microRNA expression profiles in HIV-1-transfected human cells." *Retrovirology* **2**, 81.

Yi, R., Qin, Y., Macara, I. G., and Cullen, B. R. (2003). "Exportin-5 mediates the nuclear export of pre-microRNAs and short hairpin RNAs." *Genes Dev.* **17**(24), 3011–3016.

Zuker, M. (2003). "Mfold web server for nucleic acid folding and hybridization prediction." *Nucleic Acids Res.* **31**(13), 3406–3415.

CHAPTER EIGHT

APPROACHES FOR STUDYING MICRORNA AND SMALL INTERFERING RNA METHYLATION *IN VITRO* AND *IN VIVO*

Zhiyong Yang,* Giedrius Vilkaitis,[†] Bin Yu,*
Saulius Klimašauskas,[†] *and* Xuemei Chen*

Contents

* Department of Botany and Plant Sciences and Institute of Integrative Genome Biology, University of California—Riverside, Riverside, California
† Laboratory of Biological DNA Modification, Institute of Biotechnology, Vilnius, Lithuania

Methods in Enzymology, Volume 427
ISSN 0076-6879, DOI: 10.1016/S0076-6879(07)27008-9

Abstract

The biogenesis of microRNAs (miRNAs) in plants is similar to that in animals, however, the processing of plant miRNAs consists of an additional step, the methylation of the miRNAs on the 3′ terminal nucleotides. The enzyme that methylates *Arabidopsis* miRNAs is encoded by a gene named *HEN1*, which has been shown genetically to be required for miRNA biogenesis *in vivo*. Small interfering RNAs (siRNAs) are also methylated *in vivo* in a *HEN1*-dependent manner. Our biochemical studies demonstrated that HEN1 is a methyltransferase acting on both miRNAs and siRNAs *in vitro*. HEN1 recognizes 21 to 24 nt small RNA duplexes, which are the products of Dicer-like enzymes, and transfers a methyl group from S-adenosylmethionine (SAM) to the 2′ OH of the last nucleotides of the small RNA duplexes. Here we describe methods to characterize the biochemical activities of the HEN1 protein both *in vitro* and *in vivo*, and methods to analyze the methylation status of small RNAs *in vivo*.

1. INTRODUCTION

An miRNA is a 21 to 24 nt endogenous RNA product of a non–protein-coding gene. Since the first miRNA was discovered as a regulator of developmental timing in *Caenorhabditis elegans* in 1993 (Lee *et al.*, 1993), numerous miRNAs have been identified and recognized as important regulators of gene expression in both plants and animals (reviewed in Bartel, 2004). An miRNA is derived from a long primary transcript known as pri-miRNA (Lee *et al.*, 2002). In animals, the pri-miRNA is processed by the ribonuclease III (RNase III) enzyme Drosha into a precursor miRNA (pre-miRNA) (Lee *et al.*, 2003). The pre-miRNA is subsequently processed by another RNase III enzyme, Dicer, to generate a double-stranded miRNA intermediate known as miRNA/miRNA★ (Grishok *et al.*, 2001; Hutvágner *et al.*, 2001; Ketting *et al.*, 2001; Lee *et al.*, 2003). Characteristics of the miRNA/miRNA★ duplex include 5′ P, 3′ OH, and a 2 nt 3′ overhang on each strand (Basyuk *et al.*, 2003; Lee *et al.*, 2003). In plants, both pri-miRNAs and pre-miRNAs are processed by DCL1, a dicer homolog (Kurihara and Watanabe, 2004; Park *et al.*, 2002; Reihart *et al.*, 2002). After the miRNA/miRNA★ duplex is generated, the miRNA strand of the duplex is loaded into an RNA-induced silencing complex (RISC) where it targets mRNAs through sequence complementarity to lead to their cleavage or translational repression (reviewed in Bartel, 2004).

Our work revealed another player in miRNA biogenesis in plants—HEN1. HEN1 was first identified in a genetic screen as a gene important for reproductive organ identity (Chen *et al.*, 2002). Because many aspects of the developmental defects in the *hen1-1* mutant were similar to that of *caf-1,* a mutant of *DCL1* (Chen *et al.*, 2002; Jacobsen *et al.*, 1999), we tested whether HEN1 is required for miRNA accumulation *in vivo*. Indeed, the abundance of miRNAs is dramatically reduced in *hen1* mutants, indicating that HEN1 is required for miRNA biogenesis (Park *et al.*, 2002).

We went on to demonstrate that HEN1 is an miRNA methyltransferase. The 942 aa HEN1 protein contains a conserved SAM-binding motif in the C-terminal region and a putative double-stranded RNA binding motif in the N-terminal region. The SAM-binding motif is embedded within a larger domain showing sequence similarities to many methyltransferases (Tkaczuk *et al.*, 2006; Yu *et al.*, 2005), suggesting that HEN1 is a methyltransferase. Using purified recombinant GST-HEN1 protein, we demonstrated that HEN1 is a miRNA methyltransferase that acts on miRNA/miRNA★ in a sequence-independent manner (Yu *et al.*, 2005). HEN1 can methylate small RNA duplexes ranging from 19 to 26 nt in size but has the best efficiency on 21 to 24 nt small RNA duplexes (Yang *et al.*, 2006). In addition, it has a strict requirement for the 2 nt 3′ overhang and for the presence of both the 2′ and 3′ OH on the ribose of the 3′ terminal nucleotide, characteristic features of Dicer products (Yang *et al.*, 2006).

The requirement of HEN1 for miRNA biogenesis *in vivo* and the *in vitro* biochemical activities of HEN1 suggest that plant miRNAs are methylated *in vivo*. We have demonstrated that this is indeed the case—plant miRNAs carry a methyl group on the 3′ terminal nucleotides (Yu *et al.*, 2005). Another class of small RNAs similar to miRNAs in structure and in their biogenesis is siRNAs. In contrast to miRNAs that are derived from single-stranded precursors, siRNAs are generated by Dicer-like proteins from long double-stranded RNAs. Several types of endogenous siRNAs have been found in *Arabidopsis,* such as trans–acting siRNAs, nat-siRNAs, and heterochromatic siRNAs (reviewed in Brodersen and Voinnet, 2006; Vaucheret, 2006). In addition to the endogenous siRNAs, viruses and transgenes also give rise to siRNAs (reviewed in Brodersen and Voinnet, 2006; Vaucheret, 2006). We and others have shown that all types of siRNAs (except nat-siRNAs that have not been tested) are methylated *in vivo* in a HEN1-dependent manner (Akbergenov *et al.,* 2006; Ebhardt *et al.*, 2005; Li *et al.*, 2005). Our biochemical studies with recombinant HEN1 protein also demonstrated that HEN1 is a siRNA methyltransferase (Yang *et al.*, 2006).

In this chapter, we describe molecular and biochemical methods used to delineate the activity of HEN1 as a small RNA methyltransferase *in vitro*. We report an enzymatic assay with HEN1 immunoprecipitated from plant extracts to confirm its *in vivo* activity. We also describe methods used to determine the methylation status of small RNAs *in vivo*.

2. Expression and Purification of Recombinant HEN1 Proteins

2.1. Generation of expression vectors

A modified version of the HEN1 cDNA (such that a hotspot for *Escherichia coli* transposon insertion was mutated [Yu *et al.*, 2005]) was used to generate fusion protein constructs. The HEN1 coding region was amplified with primers HEN1p4 (5′- ccg*gaattc*tcaaagatcagtctttttcttttctacatcttcttt-3′; the introduced EcoRI site is in italics) and HEN1p6 (5′- ccg*gaattc*a-tatggccggtggt-gggaagc-3′; the introduced EcoRI site is in italics) and cloned as an EcoRI fragment into pGEX-2TK for GST-HEN1 fusion protein expression in *E. coli*.

To fuse HEN1 with a shorter N-terminal peptide (MGSSHHHHHH-SSGLVPRGSH) containing a His_6-tag followed by a thrombin cleavage site, the *HEN1* cDNA was subcloned into the pET-15b vector plasmid (Novagen) in two steps. First, the 1.7-kb NdeI fragment from pGEX-2TK-HEN1 was cloned in the vector plasmid through the NdeI site. In the next step, a full-length *HEN1* gene was reconstituted by replacing the Bsu15I-XhoI fragment of the resulting plasmid with the 2.2-kb Bsu15I-EcoRI fragment from pGEX-2TK-HEN1 (the XhoI and EcoRI sites were blunt-ended by T4 DNA polymerase fill-in treatment prior to ligation).

2.2. Expression and purification of GST-tagged HEN1 protein

The *E. coli* strain BL21-CodonPlus(DE3)-RIL (Stratagene) was transformed with pGEX-2TK-HEN1. A single colony of the transformed *E. coli* was inoculated into 5 ml LB with 100 μg/ml ampicillin and 30 μg/ml chloramphenicol. After 16 h at 37° with vigorous shaking at 250 rpm, the 5-ml culture was transferred into 500 ml 2×YT medium containing 100 μg/ml ampicillin. Incubation was continued at 37° with vigorous shaking until the OD_{600} reached 0.6 to 0.9. The expression of the GST-HEN1 protein was induced by adding 500 μl of 1 M isopropyl-β-D-thiogalactopyranoside (IPTG) so that the final concentration of IPTG was 1 mM. The culture was then grown at 30° with shaking at 250 rpm for 3 to 4 h. Cells were pelleted by centrifugation at 5000g for 10 min at 4°. The supernatant was discarded and the cell pellet (kept on ice) was resuspended in approximately 20 ml of cold PBS Plus buffer (10 mM Na_2HPO_4, 1.8 mM KH_2PO4, 140-mM NaCl, 2.7 mM KCl, 1 mM DTT, 1× Protease Inhibitor Tablet without EDTA (Roche), pH 7.5). The resuspended cells were disrupted using a sonicator with a microtip (Branson 450; output at 30%). Six 10-sec sonications were performed with a 1-min interval on ice between each sonication. Twenty percent Triton X-100 was added to the sonicated solution to reach a 0.6% final concentration and mixed gently. The slurry was centrifuged at 17,000g for 20 min at 4°. The supernatant was transferred to a new tube and centrifuged again for another 10 min. The second 17,000g supernatant was utilized for protein purification with a column containing pre-prepared Glutathione Sepharose 4B Matrix (Amersham Pharmacia Biosciences). Glutathione Sepharose 4B beads were washed with PBS buffer three times to remove the 20% ethanol storage solution. Then 1-ml beads were packed into a disposable polypropylene column. The $17,000g$ supernatant was transferred into the column and allowed to slowly pass through the beads. The beads were washed with at least 10 bed vol of PBS. The GST-HEN1 protein was eluted with 10-mM reduced glutathione in 50 mM Tris-HCl buffer (pH 8.0). One-ml fractions were collected. Protein concentrations were determined using the Bradford reagent (BioRad) with BSA as a standard.

2.3. Expression and preparation of His-tagged HEN1

The protease-deficient *E. coli* strain BL21-CodonPlus(DE3)-RIL (Stratagene) was transformed with the recombinant pET-15b-HEN1 plasmid, plated on LB agar containing 100 μg/ml ampicillin and 30-μg/ml chloramphenicol, and incubated at 37°. A single colony was inoculated into 5 ml of liquid LB with the antibiotics and incubated in a shaker overnight at 37°. Five hundred ml of LB medium containing 100 μg/ml ampicillin and 30 μg/ml chloramphenicol was inoculated with a 0.005 to 0.0025 vol of the overnight culture and was cultivated at 37° until it reached a density of OD$_{600}$ = 0.6 to 0.8. Protein expression was induced by adding IPTG to a final concentration of 0.1 mM followed by incubation at 16° overnight. Cells were harvested by centrifugation at $4000g$ for 10 min (the pellet can be stored at $-70°$ for future use). The pellet was washed with ice-cold PBS, resuspended in 25 ml of Phosphate Buffer (20 mM Na$_2$HPO$_4$, 500 mM NaCl, pH 7.5) supplemented with 0.5% (v/v) Triton X-100, 5 mM 2-mercaptoethanol, and protease inhibitors (500 μM PMSF, 10 μM E-64, 1 μM pepstatin A), and sonicated. Lysed cells were centrifuged at $39,000g$ for 20 min at 4°, and the supernatant was collected and passed through a 0.45 μm pore filter. The hexahistidine tagged HEN1 protein was purified on a nickel-loaded agarose resin (Amersham Pharmacia Biosciences) according to manufacturer's recommendations. Briefly, after loading the lysate, the column was washed with Phosphate Buffer and then with 0.01-M imidazole/Phosphate Buffer until no UV absorbing material was detectable in the eluate. The desired protein was eluted with a 0.01 to 0.5 M concentration gradient of imidazole in PBS. All purification steps were performed at 4° or on ice. Purest fractions were pooled and dialyzed twice against Storage Buffer [20 mM Tris-HCl, 50 mM NaCl, 0.01% (v/v) Triton X-100, 2 mM DTT, pH 7.5] containing 5% glycerol, then against Storage Buffer containing 50% glycerol, and stored at $-20°$. The concentration and purity of the proteins were assessed using SDS-PAGE analysis (Fig. 8.1) or a colorimetric Bradford assay. The yield of the HEN1 protein obtained from 500 ml of bacterial culture was 2.5 to 5 mg.

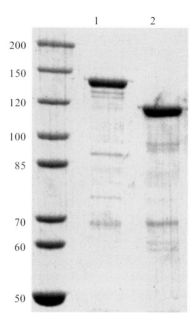

Figure 8.1 SDS-polyacrylamide gel electrophoresis of affinity-purified recombinant GST-HEN1 (lane 1) and His$_6$-HEN1 (lane 2). Protein staining with Coomassie Brilliant Blue R-250 indicates a greater than 90% purity of both recombinant methyltransferases. The sizes of molecular mass markers are indicated on the left.

 ## 3. Small RNA Methyltransferase Assays with Recombinant HEN1 Proteins

3.1. Methyltransferase assays as monitored by the incorporation of [^{14}C]-methyl groups

Single-stranded RNA oligonucleotides corresponding to miR173, miR173★, or other miRNAs or miRNA★s were purchased from Integrated DNA Technologies, Dharmacon RNA Technologies, or Metabion in high-performance liquid chromatography (HPLC) purified forms. The RNA oligonucleotides were dissolved in the annealing buffer (50 mM Tris/HCl, pH 7.6, 100 mM KCl, and 2.5 mM MgCl$_2$). The concentrations of the RNAs were determined by UV spectrometry. Equal molar amounts of the miRNA and miRNA★ strands were mixed together. Annealing was performed by heating the mixture for 5 min at 95°, slowly cooling it to 37°, followed by incubation for 2 h at 37°, and 1 h at room temperature in a thermal cycler.

A 100-μl methyltransferase reaction was set up for each pair of annealed RNAs. The reaction mixture contained 50 mM Tris-HCl (pH 8.0), 100 mM KCl, 5 mM MgCl$_2$, 0.1 mM EDTA, 2 mM DTT, 5% glycerol,

2 μl RNasin (40 U/μl; Promega), 0.5 μCi S-adenosyl-L-[methyl-^{14}C] methionine ([^{14}C]-SAM) (58.0 mCi/mmol; Amersham Pharmacia Biosciences), 5 μg purified protein, and 1 nmol RNA substrate. After incubation at 37° for 2 h, the reaction was stopped by adding 100-μl 2× proteinase K solution (100-mM Tris-HCl, pH8.0, 10-mM EDTA, 150-mM NaCl, 2% SDS, and 0.4-mg/ml proteinase K) followed by incubation at 65° for 15 min. The reaction was then extracted with phenol/chloroform. To precipitate the small RNAs, 1 μl glycogen (RNase-free, 5 mg/ml), 0.1 vol of 3-M NaOAc (pH 5.2), and 2.5 vol of ice-cold 100% ethanol were added to the reaction. The mixture was stored at −80° for 2 h and centrifuged at 4° for 30 min. The pellet was washed with 100-μl 70% cold ethanol. The RNAs in the pellet were dissolved with 1×RNA loading buffer, heated at 95° for 5 min, immediately put on ice, and loaded onto a 15% polyacrylamide gel with 7-M urea. After electrophoresis, the gel was treated with an autoradiography enhancer (En^{3}hance from Perkin Elmer) following manufacturer's instructions and exposed to X-ray film at −80°. Because the reaction results in the transfer of the ^{14}C-methyl group from SAM to the miRNAs, the miRNAs become ^{14}C-labeled after the reaction (Yu et al., 2005).

3.2. Methyltransferase assays as monitored by the incorporation of [^{3}H]-methyl groups

[methyl-^{3}H]-SAM can also be employed as the cofactor in the methylation reaction, and the enzymatically transferred methyl groups can be quantified in a liquid scintillation counter. As compared to ^{14}C label, tritium offers a 1000-fold higher specific activity and thus a higher sensitivity of detection in a scintillation counter. Therefore, the accessible concentration range can be expanded down by 2 to 3 orders of magnitude. This method is well suited for monitoring the reaction in a time-dependent manner for kinetic studies of methyltransferases (Vilkaitis et al., 2001).

Enzymatic methylation velocities were monitored by measuring the incorporation of tritiated methyl groups from [methyl-^{3}H]-SAM (Amersham Pharmacia Biosciences) onto the miR173/miR173★ duplex in the presence of HEN1. For determining steady-state velocities, methylation reactions were carried out at 37° for 60 min in 25 μl of Reaction Buffer (10 mM Tris-HCl, pH 7.5, 50 mM NaCl, 10 mM MgCl$_2$, 2 mM DTT, 0.1 mg/ml bovine serum albumin, 0.1 U/μl RiboLock^{TM} ribonuclease inhibitor [Fermentas Life Sciences, added immediately before use]), which contained 0.02 to 0.1 μM HEN1, 1 to 2 μM RNA duplex, and 20 μM [methyl-^{3}H]-SAM. Commercial [methyl-^{3}H]-SAM with two different specific activities can be purchased (0.5 Ci/mmol at 0.5 mCi/ml or 15 Ci/mmol at 1 mCi/ml). If desired, 4.7 Ci/mmol [methyl-^{3}H]-SAM could be prepared by mixing the two at a 1:6(v/v) ratio. The reactions were quenched by adding a 100-fold or higher excess of cold SAM (Sigma) and proteinase K (Fermentas Life Sciences) to a final concentration of 1 mg/ml in Stop Buffer ($2\times$ buffer: 20 mM Tris-HCl, pH 7.8, 1% SDS, 20 mM EDTA, 20 mM NaCl; store at room temperature). After 15-min incubation at $55°$, duplicate or triplicate samples were spotted onto 2.3-cm DE-81 filters (Whatman), washed four times with 0.05-M sodium phosphate buffer, pH 7.0, then two times with water, two times with ethanol, and finally once with acetone. Carefully dried filters were counted for radioactivity in 4 ml of scintillation cocktail (Rotiszint®eco plus, Carl Roth GmbH; CytoScint™, Fisher Biotech, or equivalent) using a liquid scintillation spectrometer. Background counts obtained from samples lacking dsRNA were subtracted. Figure 8.2 shows an example of the reaction time course analysis for the His$_6$-HEN1 methyltransferase. The method described above can be used for kinetic analyses of HEN1 and its derivatives.

In cases when accurate absolute numbers are required or when a significant variation in repetitive experiments is observed, it may be helpful to monitor (and to normalize the tritium counts for) the extent of recovery of the RNA duplex on DE-81 filters during the multiple wash procedures (typically 50 to 90%) (Vilkaitis *et al.*, 2001). This is achieved by adding small amounts of high specific activity ^{32}P-labeled dsRNA ($\sim$1000 dpm/sample). Because the energy spectra for the isotopes are very different, they can be measured independently using appropriate settings for spectral windows (0 to 400 for ^{3}H and 600 to 1000 for ^{32}P in Beckman scintillation counters) (Yang *et al.*, 2003).

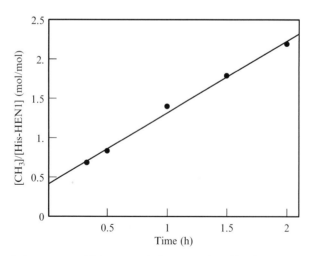

Figure 8.2   A time course of dsRNA methylation catalyzed by the His$_6$-HEN1 methyltransferase. Experiment was performed at $37°$ in Reaction Buffer containing 2-μM miR173/miR173$\star$ RNA, 20-μM [*methyl*-^{3}H]-SAM (4.7 Ci/mmol), and 0.1-μM methyltransferase. Duplicate aliquots were removed for analysis at specified time points and processed as described.

3.3. Methyltransferase assays as monitored by β-elimination

The methyltransferase reactions can also be monitored by the appearance of the products in the absence of radioactivity. The products can be distinguished from the substrates because the products, not the substrates, are resistant to sodium periodate treatment followed by β-elimination. Periodate is specific for cis-diols, which are only found in the 3′-terminal nucleotide when both of its adjacent 2′ and 3′ hydroxyl groups are unmodified. Sodium periodate (oxidation reaction) cleaves the vicinal hydroxyl groups to a dialdehyde. Then the dialdehyde is treated with borate at pH 9.5 (β-elimination reaction) to result in a 1-nt shorter RNA with a 3′-monophosphate (Alefelder et al., 1998). The RNA will move faster in gel electrophoresis by approximately 1.5 nt than the RNA prior to the treatment. Figure 8.3 shows the difference in electrophoretic mobility of an in vitro synthesized RNA oligonucleotide with two free hydroxyl groups on the terminal ribose. The product of HEN1-catalyzed reaction has a 2′-O-methyl group on the ribose of the last nucleotide (Yang et al., 2006) and is therefore resistant to the chemical treatments. Monitoring the products this way allows the methyltransferase reactions to be performed in the presence of cold SAM or no added SAM (to assay SAM copurification with recombinant HEN1).

After the methyltransferase reactions (as described above), one pmol of miRNAs was dissolved in 7.5 μl of DEPC-treated H_2O and mixed with 10 μl 2× borax/boric acid buffer (0.12 M, pH 8.6). Then 2.5-μl sodium periodate (200 mM) was added. The reaction was incubated in the dark at room temperature for 30 min. Two-μl 100% glycerol was added to the reaction and incubation was continued in the dark for 10 min at room

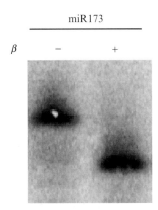

Figure 8.3 β-elimination of an RNA oligonucleotide with two free OH groups on the 3′ terminal ribose. The RNA was treated (+) or untreated (−) with the chemicals for β-elimination, resolved on a 15% polyacrylamide gel with urea, blotted to a membrane, and hybridized with an antisense probe.

temperature. The miRNAs were then precipitated and dissolved in 20 μl 1× borax/boric acid buffer (0.06 M, pH 9.5), incubated for 90 min at 45°, and then precipitated again. The pellet was washed with ice-cold 70% ethanol, dissolved in 10 μl 1× RNA loading buffer, and analyzed by Northern blotting (see following).

This assay can also be used for the quantitative detection of endogenous SAM in HEN1 preparations. The cofactor SAM is often strongly associated with methyltransferases and may be retained in preparations even after several steps of chromatographic purification (Kumar *et al.*, 1992). Substantial amounts of bound SAM (over 10 to 20 mol%) may interfere with certain biochemical and kinetic experiments. Several rounds of dialysis against a suitable buffer are usually sufficient to remove most of the bound cofactor. The detection of endogenous SAM is based on *in vitro* enzymatic reactions, in which increasing amounts of a methyltransferase are incubated with a fixed amount of its substrate without added SAM (Kumar *et al.*, 1992). For example, methylation reactions can be performed with 0.2 μM miR173/miR173* duplex and several concentrations of GST-HEN1 in the range 0.2 to 3 μM. Reactions are incubated and analyzed using the β-elimination assay as described previously to determine the concentration of product [Product] formed. Methyltransferase concentrations [Methyltransferase] that give $0.1 <$ [Product]/[Substrate] < 0.9 are selected to calculate the mol% of bound SAM according to SAM% = [Product]/[Methyltransferase] × 100%.

4. Reverse-Phase HPLC Analysis to Determine the Position of the Methyl Group in Products of HEN1-Catalyzed Reactions

Using the aforementioned β-elimination reactions, we showed that at least one of the two OH groups on the 3' end of small RNAs is blocked after methylation, suggesting that a methyl group is on either the 2' OH, the 3' OH, or both positions (Yu *et al.*, 2005). One way to determine the position of the methyl group is to analyze the terminal nucleoside by HPLC. To simplify our analysis, we designed an RNA oligonucleotide, miR173*C, by changing the last nucleotide of miR173* from G to C so that the terminal nucleotides of both strands in the miR173/miR173*C duplex are C. Because both strands are methylated, the G-to-C change ensures that only methylated cytidine be present after the reaction.

We annealed miR173 to miR173*C and performed the GST-HEN1 methylation reaction and a control reaction in which GST instead of GST-HEN1 was included. Approximately 200 μg of annealed miR173/miR173*C duplex was incubated with purified GST-HEN1 or GST in the presence of 1-mM cold SAM at 37° for 2 h. After the reaction, the miRNAs were extracted and precipitated, as described earlier.

The precipitated miR173/miR173*C duplex was dissolved in a 45-μl sodium acetate buffer (20 mM; pH 5.3) containing 5 mM ZnCl$_2$ and 50 mM NaCl, and then digested with 5 U of nuclease P1 (1000 U/ml; USB) for 60 min at 37°. Nuclease P1 is specific for single-stranded RNA and is therefore expected to fully digest the 2 nt 3′ overhang in the duplex. After digestion, 10 μl 1-M Tris-HCl (pH 8.0) was added to bring the pH back to 8.0. Then 1 U of calf intestine alkaline phosphatase (Roche) was added, and the reaction was allowed to proceed for 30 min at 37°. Following the dephosphorylation reaction, all of the miR173/miR173*C hydrolysate was subjected to reverse-phase HPLC with a Phenomenex Luna C18 (250 × 4.60 mm) column at a flow rate of 0.8 ml/min. The mobile phase was 50-mM triethylamine acetate (pH 7.6) and 2% acetonitrile (ACN). A gradient was used in which the concentration of ACN was gradually increased starting at 15 min. The program was as follows: 0 to 15 min: 2% ACN; 15 to 20 min: linear increase from 2 to 100% ACN; 20 to 25 min: 100% ACN; 25 to 30 min: linear decrease from 100 to 2% ACN. This HPLC scheme allowed the separation of 2′-O-methyl C from 3′-O-methyl C as well as from other nucleosides. Our analyses showed that HEN1-catalyzed methylation of miRNAs occurs on the 2′ OH on the ribose of the 3′ terminal nucleotides (Yang *et al.*, 2006).

5. Immunoprecipitation and HEN1 Activity Assay

5.1. 35S::HA-HEN1 construction and plant transformation

A *HEN1* genomic fragment encompassing the entire coding region plus introns was cloned into the binary plant transformation vector pPZP211 to generate pPZP211-HEN1. Then a cassette containing the 35S promoter, a translational leader, and a 3× HA epitope was cloned into pPZP211-HEN1, resulting in a translational fusion of 3× HA to the N-terminus of HEN1. This construct, when introduced into the *hen1-1* background, was able to rescue the *hen1-1* morphological defects, suggesting HA-HEN1 fusion was functional. Western blots on total protein extracts from *35S::HA-HEN1 hen1-1* inflorescences confirmed the accumulation of HA-HEN1 (data not shown).

5.2. HA-HEN1 immunoprecipitation and enzymatic activity assay

Inflorescences (5 g) from *35S::HA-HEN1 hen1-1* or the *hen1-1* control *Arabidopsis* were ground in liquid nitrogen to a fine powder. The powder was added to 30 ml IP buffer (40 mM HEPES/KOH, pH 7.4, 150 mM NaCl, 10 mM KCl, 5 mM MgCl$_2$, 1 mM EDTA, 2 mM DTT, 0.1% Triton X-100, 1 mM PMSF, 1× Complete Protease Inhibitor without EDTA

[Roche]). The slurry was filtered through three layers of miracloth (Cal-biochem), incubated on a rotating wheel at 4° for 30 min, and then centrifuged for 30 min at 17,000*g* at 4°. The supernatant was transferred to a new centrifuge tube and spun for another 20 min. The supernatant was again transferred to a new tube. A 20-μl fraction was removed from the supernatant, mixed with 20 μl 2× SDS sample buffer, and boiled for 5 min so that it served as a total input control. One hundred-μl anti-HA affinity matrix (Roche) was added to the rest of the supernatant. The mixture was incubated on a rotating wheel for 3 h at 4°. The beads were precipitated by spinning at 500 *g* for 5 min and were washed three times with PBS containing 1× Protease Inhibitor Tablet. The beads were then equilibrated with methyltransferase assay buffer (50 m*M* Tris-HCl, pH 8.0, 100 m*M* KCl, 5 m*M* MgCl$_2$, 0.1 m*M* EDTA, 5% glycerol, and 1 m*M* DTT). For the methyltransferase assay, 50 μl of beads was used for each assay using the same method as described previously (with ^{14}C incorporation as the measure for activity). For SDS-PAGE, the affinity-bound proteins were eluted from the beads by boiling for 5 min in 2× SDS sample buffer before loading onto a 10% SDS-polyacrylamide gel (Fig. 8.4A). As shown in Fig. 8.4B, the

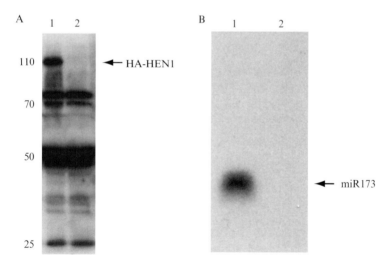

Figure 8.4 Immunoprecipitation (IP) of HEN1 followed by a methyltransferase assay. IP was performed from inflorescence tissues of *35S::HA-HEN1 hen1–1* (lane 1) or *hen1–1* (lane 2). A, Western blotting with anti-HEN1 polyclonal antibodies. Ten-μl beads from the IP were mixed with 10-μl 2× SDS sample buffer, boiled, and loaded on a 10% SDS-polyacrylamide gel for Western blotting with anti-HEN1 polyclonal antibodies. Sizes of molecular mass markers are indicated on the left. B, A HEN1 methyltransferase assay with the immunoprecipitates from *35S::HA-HEN1 hen1–1* (lane 1) and *hen1–1* (lane 2). The miR173/miR173★ duplex was used as the substrate and [*methyl*-^{14}C] SAM was used as the cofactor.

HA-HEN1 immunoprecipitate from plant extracts was able to methylate the miR173/miR173★ duplex. This confirms that HEN1 has miRNA methyltransferase activity *in vivo*.

6. ANALYSIS OF THE *IN VIVO* METHYLATION STATUS OF MiRNAS AND SiRNAS

We developed two methods to evaluate the methylation status of miRNAs and siRNAs *in vivo*. First, we combined β-elimination of total plant RNAs with small RNA Northern blots to show that plant miRNAs lack at least one free OH on the 3′ terminal nucleotides. Second, we isolated sufficient amounts of miR173 by affinity purification and subjected the purified miR173 to mass spectrometry to determine its molecular weight. The latter approach showed that miR173 is approximately 14 Daltons larger than expected, consistent with its carrying a methyl group (Yu *et al.*, 2005). Although the second method is more definitive in the determination of the methylation status of *in vivo* small RNAs, it cannot be routinely used due to the requirement for large amounts of purified RNAs. In addition, it is also limited to those small RNAs without related species *in vivo* because the affinity purification step cannot effectively distinguish related species, and the presence of related species complicates the mass spectrometry analysis. Since this approach cannot be routinely used, it is not discussed further here. Instead, we present a detailed protocol for the first method (β-elimination followed by Northern blotting).

6.1. Periodate treatment and β-elimination

Total RNA was extracted from liquid nitrogen–frozen plant tissues using Tri Reagent (Molecular Research Center, Inc.) according to manufacturer's instructions. Periodate treatment and β-elimination were performed as previously described (Alefelder *et al.*, 1998) with some modifications. Approximately 50 μg of total RNA was dissolved in 37.5 μl DEPC-treated water and mixed with 50-μl 2× borax/boric acid buffer (0.12 M, pH 8.6). Then 12.5 μl sodium periodate (200 mM in water) was added. The solution was mixed quickly and incubated in the dark at room temperature for 30 min. One hundred percent glycerol (10 μl) was added to the reaction to quench unreacted sodium periodate by incubation for an additional 10 min in the dark at room temperature. The RNA was precipitated and dissolved in 100 μl 1× borax/boric acid buffer (0.06 M, pH 9.5), incubated for 90 min at 45°, and then precipitated with ethanol. The pellet was washed with 70% ethanol and dissolved in 10 μl 1× RNA loading buffer. The treated total RNA was analyzed by small RNA Northern blotting.

6.2. Small RNA Northern blotting

The treated total RNA and control total RNA without periodate treatment were loaded on a 15% polyacrylamide gel containing 7-M urea. The RNAs were transferred to Zeta-probe GT membranes (BioRad) using semi-dry transfer equipment (Owl Separation Systems) for 1 h at 10 V. After electro-blotting, RNAs were fixed to the membrane by ultraviolet (UV) cross-linking for 1 min followed by baking in a vacuum oven at 80° for 1 h. The membrane was prehybridized with Ultrahyb-oligo hybridization buffer (Ambion) for 2 h at 40°. During the prehybridization, a 5′ end-labeled probe was prepared as follows: 5 μl 10× T4 DNA polynucleotide kinase buffer (New England Biolabs), 5 μl T4 DNA polynucleotide kinase (New England Biolabs), 0.5 μl 100-μM DNA oligonucleotide (antisense strand), 5 μl gamma [^{32}P]-ATP (6000 Ci/mmol; PerkinElmer), and 34.5 μl H$_2$O were mixed and incubated at 37° for 1 h. The oligonucleotide was isolated from the free [^{32}P]-ATP by passing the reaction mixture through a MinSpin G-25 Column (Amersham Pharmacia Biosciences). The membrane was hybridized with the 5′ end-labeled antisense oligonucleotide probe for approximately 20 h at 40°. Then the membrane was washed three times with 2× SSC/0.5% SDS at 40°. The radioactive signals were visualized and quantified with a PhosphorImager.

The expected results are such that a shift in mobility by about 1.5 nt should be observed when a small RNA is not methylated and that no shift be observed when a small RNA is methylated. However, it should be noted that an *in vitro* synthesized oligonucleotide without methylation should be spiked into total RNA as a control for the completeness of the chemical reactions. In addition, resistance to the chemical reactions only indicates that one of the two OH groups is blocked without reference to the nature of the modification on the small RNAs.

7. Concluding Remarks

The biogenesis of miRNAs and siRNAs in plants involves a critical methylation step catalyzed by the methyltransferase HEN1. HEN1 uses small RNA duplexes, products of Dicer-mediated processing of precursors molecules, as its substrates. HEN1 introduces one methyl group onto the 2′ OH of the 3′ terminal nucleotide on each strand of the duplex. *In vivo*, all miRNAs and siRNAs in *Arabidopsis* are methylated in a HEN1-dependent manner.

ACKNOWLEDGMENTS

We thank Dr. Yon W. Ebright for her help with the HPLC analysis, Drs. Manu Agarwal and Vanitharani Ramachandran for comments on the manuscript, and Alexandra Plotnikova for technical assistance. This work was funded by a National Science Foundation grant

(MCB 0343480) to X. C. The work of G. V. was supported by an EU Centers of Excellence program grant (QLK3-CT2002-30575).

REFERENCES

Akbergenov, R., Si-Ammour, A., Blevins, T., Amin, I., Kutter, C., Vanderschuren, H., Zhang, P., Gruissem, W., Meins, F., Hohn, J. T., and Pooggin, M. M. (2006). Molecular characterization of geminivirus-derived small RNAs in different plant species. *Nucleic Acids Res.* **34,** 436–444.

Alefelder, S., Patel, B. K., and Eckstein, F. (1998). Incorporation of terminal phosphorothioates into oligonucleotides. *Nucleic Acids Res.* **26,** 4983–4988.

Bartel, D. P. (2004). MicroRNAs: Genomics, biogenesis, mechanism, and function. *Cell* **116,** 281–297.

Basyuk, E., Suavet, F., Doglio, A., Bordonne, R., and Bertrand, E. (2003). Human *let-7* stem-loop precursors harbor features of RNase III cleavage products. *Nucleic Acids Res.* **31,** 6593–6597.

Brodersen, P., and Voinnet, O. (2006). The diversity of RNA silencing pathways in plants. *Trends Genet.* **22,** 268–280.

Chen, X., Liu, J., Cheng, Y., and Jia, D. (2002). *HEN1* functions pleiotropically in *Arabidopsis* development and acts in C function in the flower. *Development* **129,** 1085–1094.

Ebhardt, H. A., Thi, E. P., Wang, M. B., and Unrau, P. J. (2005). Extensive 3′ modification of plant small RNAs is modulated by helper component-proteinase expression. *Proc. Natl. Acad. Sci. USA* **102,** 13398–13403.

Grishok, A., Pasquinelli, A. E., Conte, D., Li, N., Parrish, S., Ha, I., Baillie, D. L., Fire, A., Ruvkun, G., and Mello, C. C. (2001). Genes and mechanisms related to RNA interference regulate expression of the small temporal RNAs that control *C. elegans* developmental timing. *Cell* **106,** 23–34.

Hutvágner, G., McLachlan, J., Pasquinelli, A. E., Balint, É., Tuschl, T., and Zamore, P. D. (2001). A cellular function for the RNA-interference enzyme Dicer in the maturation of the *let-7* small temporal RNA. *Science* **293,** 834–838.

Jacobsen, S. E., Running, M., and Meyerowitz, E. M. (1999). Disruption of an RNA helicase/RNAse III gene in *Arabidopsis* causes unregulated cell division in floral meristems. *Development* **126,** 5231–5243.

Ketting, R. F., Fischer, S. E., Bernstein, E., Sijen, T., Hannon, G. J., and Plasterk, R. H. (2001). Dicer functions in RNA interference and in synthesis of small RNA involved in developmental timing in *C. elegans*. *Genes Dev.* **15,** 2654–2659.

Kumar, S., Cheng, X., Pflugrath, J. W., and Roberts, R. J. (1992). Purification, crystallization, and preliminary X-ray diffraction analysis of an M.HhaI-AdoMet complex. *Biochemistry* **31,** 8648–8653.

Kurihara, Y., and Watanabe, Y. (2004). *Arabidopsis* micro-RNA biogenesis through Dicerlike1 protein functions. *Proc. Natl. Acad. Sci. USA* **101,** 12753–12758.

Lee, R. C., Feinbaum, R. C., and Ambros, V. (1993). The *C. elegans* heterochronic gene *lin-4* encodes small RNAs with antisense complementarity to *lin-14*. *Cell* **75,** 843–854.

Lee, Y., Ahn, C., Han, J., Choi, H., Kim, J., Yim, J., Lee, J., Provost, P., Radmark, O., Kim, S., and Kim, V. N. (2003). The nuclear RNase III Drosha initiates microRNA processing. *Nature* **425,** 415–419.

Lee, Y., Jeon, K., Lee, J. T., Kim, S., and Kim, V. N. (2002). MicroRNA maturation: Stepwise processing and subcellular localization. *EMBO J.* **21,** 4663–4670.

Li, J., Yang, Z., Yu, B., Liu, J., and Chen, X. (2005). Methylation protects miRNAs and siRNAs from a 3′-end uridylation activity in *Arabidopsis*. *Curr. Biol.* **15,** 1501–1507.

Park, W., Li, J., Song, R., Messing, J., and Chen, X. (2002). CARPEL FACTORY, a Dicer homolog, and HEN1, a novel protein, act in microRNA metabolism in *Arabidopsis thaliana*. *Curr. Biol.* **12**, 1484–1495.

Reinhart, B. J., Weinstein, E. G., Rhoades, M. W., Bartel, B., and Bartel, D. P. (2002). MicroRNAs in plants. *Genes Dev.* **16**, 1616–1626.

Vaucheret, H. (2006). Post-transcriptional small RNA pathways in plants: Mechanisms and regulations. *Genes Dev.* **20**, 759–771.

Vilkaitis, G., Merkiene, E., Serva, S., Weinhold, E., and Klimasauskas, S. (2001). The mechanism of DNA cytosine-5 methylation: Kinetic and mutational dissection of *Hha*I methyltransferase. *J. Biol. Chem.* **276**, 20924–20934.

Tkaczuk, K. L., Obarska, A., and Bujnicki, J. M. (2006). Molecular phylogenetics and comparative modeling of HEN1, a methyltransferase involved in plant microRNA biogenesis. *BMC Evol. Biol.* **6**, 6.

Yang, J., Trakselis, M. A., Roccasecca, R. M., and Benkovic, S. J. (2003). The application of a minicircle substrate in the study of the coordinated T4 DNA replication. *J. Biol. Chem.* **278**, 49828–49838.

Yang, Z., Ebright, Y. W., Yu, B., and Chen, X. (2006). HEN1 recognizes 21–24 nt small RNA duplexes and deposits a methyl group onto the 2′ OH of the 3′ terminal nucleotide. *Nucleic Acids Res.* **34**, 667–675.

Yu, B., Yang, Z., Li, J., Minakhina, S., Yang, M., Padgett, R. W., Steward, R., and Chen, X. (2005). Methylation as a crucial step in plant microRNA biogenesis. *Science* **307**, 932–935.

CHAPTER NINE

ANALYSIS OF SMALL RNA PROFILES DURING DEVELOPMENT

Toshiaki Watanabe,*,[†] Yasushi Totoki,[‡] Hiroyuki Sasaki,*,[†] Naojiro Minami,[§] *and* Hiroshi Imai[§]

Contents

* Division of Human Genetics, Department of Integrated Genetics, National Institute of Genetics, Research Organization of Information and Systems, Mishima, Japan
[†] Department of Genetics, School of Life Science, The Graduate University for Advanced Studies (SOKENDAI), Mishima, Japan
[‡] Genome Annotation and Comparative Analysis Team, Computational and Experimental Systems Biology Group, RIKEN Genomic Sciences Center, Tsurumi-ku, Yokohama, Kanagawa, Japan
[§] Laboratory of Reproductive Biology, Department of Agriculture, Kyoto University, Kyoto, Japan

Methods in Enzymology, Volume 427
ISSN 0076-6879, DOI: 10.1016/S0076-6879(07)27009-0

Abstract

Small RNAs ranging in size between 20 and 32 nt regulate gene expression through chromatin modification, mRNA degradation, and translational repression. Three major classes of small RNAs have been characterized: microRNAs (miRNAs), short interfering RNAs (siRNAs), and Piwi-interacting RNAs (piRNAs). miRNAs are expressed in a developmentally regulated and tissue-specific manner and are involved in development and cell differentiation. siRNAs are mainly involved in defense against transposons and viruses. piRNAs are expressed in germ cells and stem cells and are thought to repress transposition of retrotransposons. In this chapter, we describe the methods of small RNA cloning, annotation and classification, and their expression analysis during development.

1. INTRODUCTION

Small RNAs are 20- to 32-nt RNA molecules that guide various processes involving sequence-specific silencing through chromatin modification, mRNA degradation, and translational repression. Three major classes (miRNAs, siRNAs, piRNAs) of small RNAs have been characterized in eukaryotes. miRNAs and siRNAs are present in diverse animals and plant (Bartel, 2004). By contrast, piRNAs (also called germline small RNAs, gsRNAs) are found only in animals (Kim, 2006). miRNAs are generated by RNaseIII Dicer enzyme from stem-loop (pre-miRNAs) molecules, and siRNAs are generated by RNaseIII Dicer enzyme from dsRNA molecules. Dicer-generated siRNAs and miRNAs are incorporated into Argonaute protein and then repress gene expression in a sequence-specific manner (Bartel, 2004). piRNAs are thought to be generated from longer precursor molecules. However, the piRNA-generating enzymes and its recognition motifs in piRNA precursors remain unknown. piRNAs are incorporated into Piwi proteins, a subfamily of Argonaute proteins and regulate gene expression (Aravin *et al.*, 2006; Girard *et al.*, 2006; Grivna *et al.*, 2006; Lau *et al.*, 2006; Watanabe *et al.*, 2006).

miRNAs are found in all multicellular organisms. In human, about 500 miRNAs have been identified (Berezikov *et al.*, 2005, 2006). Most miRNAs show tissue-specific or developmental stage-specific expression and are involved in cell differentiation and developmental transitions (Aravin *et al.*, 2003; Berezikov *et al.*, 2006; Chen *et al.*, 2005; Lagos-Quintana *et al.*, 2002; Watanabe *et al.*, 2005). Many studies have shown miRNAs play important roles in development and differentiation. For example, miR-430 in zebrafish is first expressed at the time of zygotic gene activation and is involved in the clearance of the maternal mRNA during early embryogenesis (Giraldez *et al.*, 2006). miR-196 in vertebrates targets the *Hox* genes during development and spatially restricts its expression (Yekta *et al.*, 2004). miR-1 in animals is differentially expressed in the mesoderm and muscle cells and promotes muscle proliferation and differentiation (Sokol and Ambros, 2005).

Endogenous siRNAs are classified into at least four classes (rasiRNAs, ta-siRNAs, nat-siRNAs, secondary siRNAs) (Aravin and Tuschl, 2005). Repeat-associated siRNAs (rasiRNAs) have sequences of repetitive elements such as transposons and centromeric repeats. In *Schizosaccharomyces pombe*, rasiRNAs derived from centromeric repeats initiate heterochromatin formation at the centromere regions (Volpe *et al.*, 2002). In *Arabidopsis thaliana*, rasiRNAs corresponding to transposons and other repetitive elements induce DNA methylation at corresponding loci (Chan *et al.*, 2004). In mouse, retrotransposon-derived siRNAs repress retrotransposon sequences during oogenesis and early embryogenesis (Svoboda *et al.*, 2004; Watanabe *et al.*, 2006). rasiRNAs in *Drosophila melanogaster* are longer than siRNAs found in other animals and are now considered to be another class of small RNAs (see later) (Aravin *et al.*, 2003; Vagin *et al.*, 2006). Trans-acting siRNAs (ta-siRNAs) in *A. thaliana* are generated from double-stranded RNAs synthesized by RNA-dependent RNA polymerase and direct the cleavage of target mRNAs (Allen *et al.*, 2005). nat-siRNAs in *A. thaliana* are generated from double-stranded RNAs derived from natural antisense transcripts induced by salt stress (Borsani *et al.*, 2005). Secondary siRNAs, which are produced by utilizing RNA-dependent RNA polymerase, in *C. elegans* are only of antisense polarity, carry 5′ diphosphates or triphosphates, and comprise a distinct class from primary siRNAs (Pak and Fire, 2006; Ruby *et al.*, 2006; Sijen *et al.*, 2006; Yigit *et al.*, 2006).

piRNAs are 25 to 32 nt in length and have been found in *Drosophila* and mammals. In *Drosophila*, most piRNAs are derived from repetitive regions, such as retrotransposons and heterochromatic regions (Saito *et al.*, 2006; Vagin *et al.*, 2006). However, in mammals, only ∼10% of piRNAs correspond to the repeat sequences during male meiosis. Most of the other piRNAs in mammals have been mapped to unique genomic sequences (Aravin *et al.*, 2006; Girard *et al.*, 2006; Grivna *et al.*, 2006; Watanabe *et al.*, 2006). Sequences of nonrepetitive piRNAs tend to occur in clusters with a strong strand bias, suggesting that piRNAs are not of dsRNA origin (Watanabe *et al.*, 2006). This view is consistent with the finding that the Dicer genes in *Drosophila* have no effect on piRNA accumulation (Vagin *et al.*, 2006). Repetitive piRNAs are thought to repress retrotransposons transcriptionally or posttranscriptionally, whereas the functions of nonrepetitive piRNAs, which are mapped to ∼100 clusters in the mouse genome, remain unknown (Aravin *et al.*, 2006; Girard *et al.*, 2006; Grivna *et al.*, 2006; Watanabe *et al.*, 2006).

Some species-specific small RNAs also have been found in several organisms. In *Tetrahymena thermophila*, two classes of small RNAs have been found (Lee and Collins, 2006). One class is called scan(scn) RNAs, which are 27 to 30 nt in length and are involved in DNA elimination during macronucleus formation (Mochizuki *et al.*, 2002). Biogenesis of scnRNAs requires a Dicer-like protein, suggesting that they are of dsRNA origin (Mochizuki and Gorovsky, 2005). The other class is 23 to 24 nt in length and is thought to be involved in posttranscriptional regulation. Tiny noncoding RNAs (tncRNAs) in *C. elegans* are similar in length to miRNAs, but are neither processed from hairpin-like structures nor conserved between species (Ambros *et al.*, 2003b). 21U-RNAs in *C. elegans* are precisely 21 nucleotides long, and originate from more than 5700 genomic loci. These loci share a large upstream motif (Ruby *et al.*, 2006). In this chapter, we describe the methods for preparing small RNAs from embryos or tissues, cloning small RNAs using a commercially available kit, classifying small RNAs using the databases, and analyzing the expression of small RNAs during development by Northern blotting.

2. Preparation of Low–Molecular-Weight RNA and Urea-Polyacrylamide Gel Electrophoresis (Urea-PAGE)

Small RNA cloning and Northern blotting usually require 50 μg of total RNA. Loading such a large amount of RNA on a gel results in low resolution and blurred bands. Therefore, enrichment of low–molecular-weight RNA using a polyethylene glycol (PEG) solution or ion-exchange column is highly recommended, which helps to isolate small RNAs that can then be observed more clearly on acrylamide gels (see Note). For high-resolution electrophoresis of small RNAs ranging in size between 20 and 30 nt, a 15% urea-polyacrylamide gel is recommended.

2.1. Principle

PEG solutions precipitate large-molecular-mass nucleic acid. Low–molecular-weight nucleic acids are then precipitated from the supernatant with isopropanol or ethanol.

2.2. Methods

1. Total RNA is extracted from embryos or tissues at defined developmental stages (washing step using 70 to 80% ethanol after alcohol precipitation of RNA is not needed; see Note) and dissolved in water at a final concentration of ~2 μg/μl. An equal amount of PEG solution is added to the total RNA. Mix thoroughly by vortexing and store the tube on ice for 1 h to precipitate large–molecular-weight RNAs.
2. Centrifuge the tube at 12,000g for 10 min at 4°. Transfer the supernatant to a fresh tube and add an equal amount of ice-cold isopropanol. Mix thoroughly by vortexing and store the tube at −20° for 1 h to precipitate low–molecular-weight RNAs.

3. Centrifuge the tube at 12,000g for 30 min at 4°. Remove the supernatant and briefly centrifuge to collect the residual liquid at the bottom of the tube. Remove the residual liquid and keep the tube open at room temperature for 10 min for drying.

4. Dissolve the low–molecular-weight RNA in water at a final concentration of ~2 μg/μl (~0.1 amount of total RNA is obtained). The concentration of low–molecular-weight RNA is estimated by absorbance at 260 nm. A solution with an A$_{260}$ of 1 contains ~40 μg/ml of low-molecular-weight RNA.

5. Prepare the equipment for acrylamide gel electrophoresis (minigel, 10 × 10 × 0.1 cm, 10-ml gel vol). To prepare a 15% urea-polyacrylamide gel, mix the following reagents: 0.5 ml of 20× TBE, 3.75 ml 40% acrylamide (19:1) solution, 4.2-g urea. Add water to a final volume of 10 ml and stir the mixture at room temperature (if not dissolved, warm the mixture) on magnetic stirrer until the urea dissolves. Add 50 μl of 10% ammonium persulfate and 10 μl of TEMED and mix and pour into the apparatus. The gel should polymerize within 30 min.

6. Add an equal volume of loading buffer to 2 μg of low–molecular-weight RNA. Incubate the mixture at 65° for 15 min to denature RNA and place on ice for at least 1 min.

7. Once the gel has polymerized, carefully remove the comb and wash the well using water. Place the gel plate in the apparatus. Add 0.5× TBE to the upper and lower parts of the gel unit and prerun for 5 min at 100 V. Turn off the power supply and wash the well to remove the accumulated urea.

8. Load the samples in wells and start the electrophoresis at constant voltage (100 V).

9. When the bromophenol blue dye (the faster migrating dye is bromophenol blue) reaches the bottom of the gel, turn off the power supply and stain the gel with ethidium bromide solution for 20 min. If good-quality low–molecular-weight RNA is obtained, clear bands of tRNAs and 5S rRNA appear at 70 to 120 nt (the bands can be seen above the xylene cyanol dye). An example of the results obtained is shown in Fig. 9.1.

2.3. Materials

1. Fresh embryos or tissues of specific stages; samples frozen immediately after isolation using liquid nitrogen and stored at −80° are also possible

2. Guanidium thiocyanate-phenol-based RNA extraction reagents, such as TRIzol (Invitrogen)

3. PEG solution: 13% (w/v) polyethylene glycol 6000 or 8000, 1.6 M NaCl; autoclave and store at room temperature

4. 100% isopropanol, 80% ethanol; store at −20°

5. TBE (20×): 1.78 M Tris base, 1.78 M boric acid, 40 mM EDTA (pH 8.0); autoclave and store at room temperature
6. RNase-free water
7. Formamide loading buffer: 0.5 ml 20× TBE, 10-mg bromphenol blue, 10 mg xylene cyanol, 10 ml formamide; store at 4°
8. 40% acrylamide/bis solution (38:2); store at 4°
9. Ammonium persulfate: prepare 10% solution in water and freeze in single-use aliquots (200 μl); store at −20°
10. N,N,N,N'-Tetramethyl-ethylenediamine (TEMED)
11. Urea
12. Ethidium bromide solution in water (∼2 μg/ml) for staining

2.4. Notes

To isolate low–molecular-weight RNA from samples that contain abundant sugars, the PEG precipitation method is not recommended because this method is not able to remove sugars. Electrophoresis of sugar-containing RNA results in blurred bands. An ion-exchange column can be used to obtain sugar-free low–molecular-weight RNA. Because short RNAs are

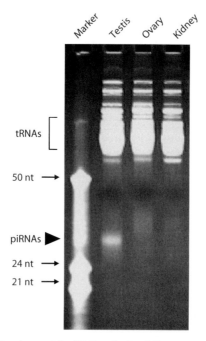

Figure 9.1 Low-molecular-weight RNAs obtained from mouse testes and other tissues. Approximately 20 μg of low–molecular-weight RNAs from each mouse tissue were loaded on a 15% polyacrylamide gel and stained with ethidium bromide. Bands of tRNAs and piRNAs are visible and depletion of high–molecular-weight RNA is confirmed.

to some extent soluble in 70 to 80% ethanol, washing is not recommended at any time when handling small RNAs.

3. Cloning of Small RNAs

Currently, a convenient small RNA cloning kit is available from Takara. We use this kit to clone small RNAs from only 1 μg of total RNA (see Note). It takes only a few days to clone small RNAs from total RNA.

3.1. Principle

Small RNAs are gel-purified, and a 3′-biotinylated linker RNA is ligated to the small RNAs. After this step, biotinylated RNA is bound to streptavidin beads. Subsequently, 5′ linker ligation and reverse transcription (RT) reaction are carried out on the beads.

3.2. Methods

1. Load the low–molecular-weight RNA (50 μg) in the wells of the two center lanes of a minigel. Load the RNA size markers in another lane. To avoid contaminating the samples with the RNA size markers, leave three lanes empty between the samples and the size markers. After loading the samples, start electrophoresis at 100 V until the bromophenol blue dye reaches the bottom of the gel.
2. Stain the gel with ethidium bromide solution for 20 min and then observe the gel under ultraviolet (UV) light. Cut out the gel fragment that contains small RNAs according to the RNA size markers and put the gel fragments into 1.5-ml tubes. Crush the gel fragments in the tubes by using pipette tips, and then add enough TE buffer to cover the gel fragments. Incubate at 55° for 1 h to elute the nucleic acids.
3. Transfer the supernatants to fresh tubes and centrifuge at 15,000 rpm for 5 min to remove the debris. Transfer the supernatants to fresh tubes again. Measure the amount of solution and add 0.1 amount of 3 M sodium acetate (pH 5.6), 1 μl of glycogen as carrier, and 3 vol of ice-cold 100% ethanol to each tube. Store the tubes at −80° for at least 1 h to precipitate the nucleic acids.
4. Centrifuge at 15,000 rpm for 30 min at 4°. Remove the supernatant and briefly centrifuge to collect the residual liquid at the bottom of the tube. Remove the residual liquid and keep the tubes open at room temperature for 10 min for drying.

5. After dephosphorylation, ligate 3′- and 5′-linkers and subsequently perform RT and PCR reaction according to the manufacturer's instructions.
6. Run 10 μl of the PCR product and a 10-bp DNA marker on a 10% nondenatured acrylamide gel at 100 V until the bromophenol blue dye reaches the bottom of the gel.
7. Stain the gel with ethidium bromide solution for 20 min and then observe the gel under UV light. At this point, two bands of approximately 43 bp (the product of directly joined 5′ and 3′ linker) and 65 bp (the product with a small RNA between linkers) should be visible on the gel. Cut the portion of the gel that contains DNA of 60 to 75 nt in length according to DNA size markers and put the gel fragment into a 1.5-ml tube. Crush the gel fragments in the tubes by using pipette tips, and then add enough TE buffer to cover the gel fragments. Incubate at 80° for 1 hr to elute the nucleic acids.
8. Transfer the supernatants to fresh tubes and centrifuge at 15,000 rpm for 5 min to remove the debris. Transfer the supernatants to fresh tubes again. Measure the amount of solution and add 0.1 amount of 3 M sodium acetate (pH 5.6), 1 μl of glycogen as carrier and 3 vol of ice-cold 100% ethanol to each tube. Store the tubes at −80° for at least 1 h to precipitate the nucleic acids.
9. Centrifuge for 30 min at 15,000 rpm at 4°. Remove the supernatant and briefly centrifuge to collect the residual liquid at the bottom of the tube. Remove the residual liquid and keep the tubes open at room temperature for 10 min for drying.
10. Add 10 μl of water to the tube and vortex.
11. Perform a second PCR reaction using a 1-μl portion of the template solution prepared at step 10 and the same primers used in the first PCR reaction. Run 10 or 15 cycles.
12. Run a 10-μl portion of the PCR product and a 10-bp DNA marker on a 10% nondenatured acrylamide gel at 100 V until the bromophenol blue dye reaches the bottom of the gel.
13. Stain the gel with ethidium bromide solution for 20 min and then observe under UV light. At this point, only a single band of approximately 65 bp should be detected on the gel.
14. Clone and sequence these fragments.

3.3. Materials

Small RNA cloning kit (Takara)
Low–molecular-weight RNA from embryos or tissues
15% urea acrylamide gel
10% nondenatured acrylamide gel
Phenol, pH 8.0

Chloroform
100% and 80% ethanol; store at $-20°$
10-bp DNA ladder marker
RNA marker

3.4. Note

If initial amount of total RNA is small (1–20 μg), the steps to enrich low–molecular-weight RNA should be avoided and the amount of linker RNA that is ligated to small RNAs should be decreased to 0.1 of the amount recommended in step 5.

4. CLASSIFICATION OF SMALL RNAS

The small RNA fractions contain diverse classes of RNAs: functional small RNAs (miRNAs, siRNAs, rasiRNAs, and piRNAs) and degradation products of other RNAs (tRNAs, rRNAs, snRNAs, snoRNAs, scRNAs, srpRNAs, and mRNAs). In some model organisms, annotation of genomic sequences has been advanced, whereas in others, the genome projects have not yet been completed. Therefore, the optimal method to classify small RNAs is different from species to species. The classification method also depends on the number of sequences to be analyzed. When the number of sequence is large, it is recommended to make pipeline programs in house to classify small RNAs. Here, we introduce two methods to classify small RNAs from mouse. One is a method using web sites for small-scale analysis (about 1000 small RNAs) and another is a method using pipeline programs developed for large-scale analysis (Fig. 9.2).

4.1. Methods

(Classification using web sites for small-scale analysis)

1. Blast small RNA sequences at NCBI using BLAST and the nr database (http://www.ncbi.nlm.nih.gov/) and at the miRNA registry (http://microrna.sanger.ac.uk/sequences/index.shtml) (see Fig. 9.2).
2. When a small RNA is not annotated in the first search, a 200 bp of genomic sequence encompassing the small RNA is used for the next searches.
3. To determine whether the small RNA is a breakdown product of tRNA, the tRNAscan-SE program (http://www.genetics.wustl.edu/eddy/tRNAscan-SE/) is used. To determine whether the small RNA is rasiRNA, the Repeat Masker program (http://www.repeatmasker.org/) is used. To determine whether the small RNA is a novel miRNA, the MFOLD 3.2 program (http://www.bioinfo.rpi.edu/applications/mfold/)

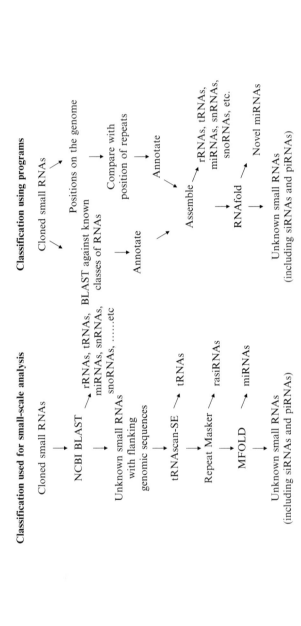

Figure 9.2 Procedure for classification of small RNAs. When the number of analyzed small RNAs is not large (100 to 1000), we annotate the small RNAs using various web servers (left). First, sequences of small RNAs are blasted using NCBI BLAST. Small RNAs not classified in this search are subjected to next searches using tRNAscan-SE, Repeat Masker, and MFOLD. When small RNAs are sequenced using a high-throughput sequencer, such as MPSS or the 454 Life Sciences system, we classify small RNAs using computers (right). Sequences of small RNAs are blasted against the database of known functional RNA sequences. At the same time, genomic positions of small RNAs are compared with that of repeats. Data of BLAST and repeats are combined. To find novel miRNAs, small RNAs not classified in these searches were analyzed by the RNAfold program.

is used (a guideline to the annotation of novel miRNA has been described in Ambros *et al.*, 2003a).

(Classification using pipeline programs for large-scale analysis)

1. After masking the linker sequences and vector sequences, sequences of small RNAs are extracted (see Fig. 9.2).
2. The extracted sequences are mapped to the genome using the BLAST program (http://www.ncbi.nlm.nih.gov/BLAST/download.shtml), and their genomic positions are retrieved.
3. To identify small RNAs corresponding to tRNAs, rRNAs, snRNAs, snoRNAs, scRNAs, srpRNAs, and rasiRNAs based on genomic position, the positions of repeats are downloaded from the University of California, Santa Cruz (UCSC) web site (http://hgdownload.cse.ucsc. edu/downloads.html). Compare the results of step 2 and the data from UCSC and classify the small RNAs.
4. To identify small RNAs corresponding to tRNAs, rRNAs, snRNAs, snoRNAs, scRNAs, miRNAs, and mRNAs based on sequence identity, extract the sequences of these RNAs from the flat files of GenBank (ftp://ftp.ncbi.nlm.nih.gov/genbank/) and download some FASTA sequences from the following databases.

   tRNAs : Genomic tRNA Database (http://lowelab.ucsc.edu/GtRNAdb/ Mmusc/)

   rRNAs : European ribosomal RNA database (http://www.psb.ugent.be/ rRNA/index.html)

   snoRNAs : snoRNA database (http://www-snorna.biotoul.fr) and RNA database (http://jsm-research.imb.uq.edu.au/rnadb/)

   miRNAs : miRBase (http://microrna.sanger.ac.uk/sequences/index. shtml)
5. Perform a BLAST search using the small RNA sequences obtained at step 1 as queries and the sequences downloaded in the step 4 as the database and classify the small RNAs.
6. Assemble the data from steps 3 and 5 and classify the small RNAs.

5. NORTHERN BLOT ANALYSIS

5.1. Principle

Because small RNAs are 20 to 30 nt in length, the optimal hybridization temperature is 40 to 50° for Northern blotting. Oligonucleotide probes labeled with $[\gamma\text{-}^{32}P]$ at their $5'$ phosphate group are optimal for this low-annealing temperature. Use of longer-transcribed ribo probes or random-primed probes at this temperature results in cross-hybridization to abundant RNA species such as rRNAs. As much as 5 μg of low–molecular-weight

RNA is required for detection of most of the abundantly expressed miRNAs. Therefore, stripping and reprobing the membrane is recommended when RNAs from rare samples are blotted.

5.2. Methods

1. Load the low–molecular-weight RNA (5–30 μg) on a 15% urea-polyacrylamide gel (2-mm thick midigel, see Note). After loading the samples, start the electrophoresis at 100 V until the bromophenol blue dye reaches the bottom of the gel.
2. Before finishing the electrophoresis, prepare a Hybond XL membrane and eight sheets of filter paper that are cut to a size just larger than the gel. The membrane and the four sheets of filter paper are then submerged in the 0.5× TBE buffer.
3. Disconnect the gel unit from the power supply and disassemble it. Lay the gel on top of the four sheets of wet filter papers, and then lay the membrane on top of the gel. Lay four additional sheets of wet filter paper on the membrane, ensuring that no bubbles are trapped in the resulting sandwich. Lay the sandwich on a transfer cassette and place it into the transfer tank so that the nylon membrane is between the gel and the anode. Transfer is accomplished at 100 mA after 2 h (wet-type apparatus).
4. After the transfer, rinse the membrane in 2× SSC for 10 min on a rocking platform and then link the RNA to the membrane using a UV cross-linker or transilluminator. The membrane can be stored for at least 1 mo at room temperature.
5. For prehybridization, insert the membrane into the hybridization bottle and then pour 10 ml of PerfectHyb hybridization buffer into the bottle. Put the bottle in the hybridizer and rotate for at least 1 h at 40°.
6. For end labeling of an oligo DNA probe that is complementary to the small RNA, mix the following reagents: 1 μl 10× polynucleotide kinase (PNK) buffer, 1 μl oligo DNA solution (10 μM), 1 μl PNK and 7 μl [γ-^{32}P] ATP, and then incubate for 30 min at 37°.
7. To prepare a size exclusion chromatography column, tamp a sterile glass wool plug into the bottom of a sterile 1-ml syringe. Put the syringe into a 15-ml tube and then fill the syringe with slurry of G-25 or G-50 Sepharose. Centrifuge for 5 min at 3000 rpm, and then put the syringe into a new 15-ml tube.
8. Add 90 μl of water to the tube containing the radiolabeled oligo DNA, and then load the radiolabeled oligo DNA solution onto the column. Centrifuge for 5 min at 3000 rpm, and then collect the eluted fraction (containing the radiolabeled oligo DNA) into a new tube. Use a liquid scintillation counter to measure the radioactivity of the eluted fraction and confirm the end labeling of oligo DNA.

9. Put the eluted fraction (radiolabeled oligo DNA) directly into the prehybridizing bottle and hybridize for 20 h at 40°.
10. Discard the hybridization solution and wash the membrane using 15 ml of low stringency buffer at 50° for 5 min. Wash three more times using 15 ml of low stringency buffer at 50° for 15 min.
11. Remove the membrane from the bottle, and then wrap it in a sheet of cellophane wrap. Put the membrane into an IP cassette and place the cassette on the imaging plate for 1 to 3 days depending on the signal intensity. Detect signals by a PhosphorImager.
12. Once the result of an miRNA expression profile has been obtained, the membrane can be stripped of the signal and then reprobed with a probe that recognizes other small RNAs. Incubate the membrane with 20 ml of stripping buffer for 30 min at 70° and then wash once in 10 ml 2× SSC at room temperature. The membrane is then ready to prehybridize.

5.3. Materials

SSC (20×): 3 M NaCl, 0.3 M trisodium citrate, adjust pH to 7.0 using 1 N HCl
Low stringency wash buffer: SSC (2×), 0.1% SDS
Hybond XL membrane (GE Healthcare)
Oligo DNA probes, which are complementary to small RNAs
T4 polynucleotide kinase (NEB)
~6000 Ci/mmol [γ-^{32}P]ATP (GE Healthcare)
PerfectHyb Plus hybridization solution (Sigma)
Sephadex G-25 or G-50 (GE Healthcare)
Stripping buffer: 10 mM Tris-Cl (pH 7.4), 0.2% SDS

5.4. Notes

The use of a thick gel (2 mm) results in sharper and clearer bands. The transfer efficiency of small RNAs in a semidry-type apparatus is usually the same as that in a wet-type apparatus. However, the wet-type apparatus transfers larger RNA species (such as pre-miRNA) more efficiently than the semidry-type apparatus. To confirm equal loading of samples from different developmental stages, load 0.1 to 0.2 amounts of RNAs that are loaded for Northern blot analysis on a 15% urea gel and stained using ethidium bromide solution.

REFERENCES

Allen, E., Xie, Z., Gustafson, A. M., and Carrington, J. C. (2005). microRNA-directed phasing during trans-acting siRNA biogenesis in plants. *Cell* **121,** 207–221.

Ambros, V., Bartel, B., Bartel, D. P., Burge, C. B., Carrington, J. C., Chen, X., Dreyfuss, G., Eddy, S. R., Griffiths-Jones, S., Marshall, M., Matzke, M., Ruvkun, G., *et al.* (2003a). A uniform system for microRNA annotation. *RNA* **9**, 277–279.

Ambros, V., Lee, R. C., Lavanway, A., Williams, P. T., and Jewell, D. (2003b). Micro-RNAs and other tiny endogenous RNAs in *C. elegans*. *Curr. Biol.* **13**, 807–818.

Aravin, A., Gaidatzis, D., Pfeffer, S., Lagos-Quintana, M., Landgraf, P., Iovino, N., Morris, P., Brownstein, M. J., Kuramochi-Miyagawa, S., Nakano, T., Chien, M., Russo, J. J., *et al.* (2006). A novel class of small RNAs bind to MILI protein in mouse testes. *Nature* **442**, 203–207.

Aravin, A, and Tuschl, T. (2005). Identification and characterization of small RNAs involved in RNA silencing. *FEBS Lett.* **579**, 5830–5840.

Aravin, A. A., Lagos-Quintana, M., Yalcin, A., Zavolan, M., Marks, D., Snyder, B., Gaasterland, T., Meyer, J., and Tuschl, T. (2003). The small RNA profile during *Drosophila melanogaster* development. *Dev. Cell* **5**, 337–350.

Bartel, D. P. (2004). MicroRNAs: Genomics, biogenesis, mechanism, and function. *Cell* **116**, 281–297.

Berezikov, E., Guryev, V., van de Belt, J., Wienholds, E., Plasterk, R. H., and Cuppen, E. (2005). Phylogenetic shadowing and computational identification of human microRNA genes. *Cell* **120**, 21–24.

Berezikov, E., Thuemmler, F., van Laake, L. W., Kondova, I., Bontrop, R., Cuppen, E., and Plasterk, R. H. (2006). Diversity of microRNAs in human and chimpanzee brain. *Nat. Genet.* **38**, 1375–1377.

Borsani, O., Zhu, J., Verslues, P. E., Sunkar, R., and Zhu, J. K. (2005). Endogenous siRNAs derived from a pair of natural cis-antisense transcripts regulate salt tolerance in *Arabidopsis*. *Cell* **123**, 1279–1291.

Chan, S. W., Zilberman, D., Xie, Z., Johansen, L. K., Carrington, J. C., and Jacobsen, S. E. (2004). RNA silencing genes control *de novo* DNA methylation. *Science* **303**, 1336.

Chen, P. Y., Manninga, H., Slanchev, K., Chien, M., Russo, J. J., Ju, J., Sheridan, R., John, B., Marks, D. S., Gaidatzis, D., Sander, C., Zavolan, M., *et al.* (2005). The developmental miRNA profiles of zebrafish as determined by small RNA cloning. *Genes Dev.* **19**, 1288–1293.

Giraldez, A. J., Mishima, Y., Rihel, J., Grocock, R. J., Van Dongen, S., Inoue, K., Enright, A. J., and Schier, A. F. (2006). Zebrafish MiR-430 promotes deadenylation and clearance of maternal mRNAs. *Science* **312**, 75–79.

Girard, A., Sachidanandam, R., Hannon, G. J., and Carmell, M. A. (2006). A germline-specific class of small RNAs binds mammalian Piwi proteins. *Nature* **442**, 199–202.

Grivna, S. T., Beyret, E., Wang, Z., and Lin, H. (2006). A novel class of small RNAs in mouse spermatogenic cells. *Genes Dev.* **20**, 1709–1714.

Kim, V. N. (2006). Small RNAs just got bigger: Piwi-interacting RNAs (piRNAs) in mammalian testes. *Genes Dev.* **20**, 1993–1997.

Lagos-Quintana, M., Rauhut, R., Yalcin, A., Meyer, J., Lendeckel, W., and Tuschl, T. (2002). Identification of tissue-specific microRNAs from mouse. *Curr. Biol.* **12**, 735–739.

Lau, N. C., Seto, A. G., Kim, J., Kuramochi-Miyagawa, S., Nakano, T., Bartel, D. P., and Kingston, R. E. (2006). Characterization of the piRNA complex from rat testes. *Science* **313**, 363–367.

Lee, S. R., and Collins, K. (2006). Two classes of endogenous small RNAs in *Tetrahymena thermophila*. *Genes Dev.* **20**, 28–33.

Mochizuki, K., Fine, N. A., Fujisawa, T., and Gorovsky, M. A. (2002). Analysis of a piwi-related gene implicates small RNAs in genome rearrangement in *Tetrahymena*. *Cell* **110**, 689–699.

Mochizuki, K., and Gorovsky, M. A. (2005). A Dicer-like protein in *Tetrahymena* has distinct functions in genome rearrangement, chromosome segregation, and meiotic prophase. *Genes Dev.* **19,** 77–89.

Pak, J., and Fire, A. (2007). Distinct Populations of Primary and Secondary Effectors During RNAi in *C. elegans*. *Science* **315,** 241–244.

Ruby, J. G., Jan, C., Player, C., Axtell, M. J., Lee, W., Nusbaum, C., Ge, H., and Bartel, D. P. (2006). Large-scale sequencing reveals 21U-RNAs and additional Micro-RNAs and endogenous siRNAs in. *C. elegans*. *Cell* **127,** 1193–1207.

Saito, K., Nishida, K. M., Mori, T., Kawamura, Y., Miyoshi, K., Nagami, T., Siomi, H., and Siomi, M. C. (2006). Specific association of Piwi with rasiRNAs derived from retrotransposon and heterochromatic regions in the *Drosophila* genome. *Genes Dev.* **20,** 2214–2222.

Sijen, T., Steiner, F. A., Thijssen, K. L., and Plasterk, R. H. (2007). Secondary siRNAs result from unprimed RNA synthesis and form a distinct class. *Science* **315,** 244–247.

Sokol, N. S., and Ambros, V. (2005). Mesodermally expressed *Drosophila* microRNA-1 is regulated by Twist and is required in muscles during larval growth. *Genes Dev.* **19,** 2343–2354.

Svoboda, P., Stein, P., Anger, M., Bernstein, E., Hannon, G. J., and Schultz, R. M. (2004). RNAi and expression of retrotransposons MuERV-L and IAP in preimplantation mouse embryos. *Dev. Biol.* **269,** 276–285.

Vagin, V. V., Sigova, A., Li, C., Seitz, H., Gvozdev, V., and Zamore, P. D. (2006). A distinct small RNA pathway silences selfish genetic elements in the germline. *Science* **313,** 320–324.

Volpe, T. A., Kidner, C., Hall, I. M., Teng, G., Grewal, S. I., and Martienssen, R. A. (2002). Regulation of heterochromatic silencing and histone H3 lysine-9 methylation by RNAi. *Science* **297,** 1833–1837.

Watanabe, T., Takeda, A., Mise, K., Okuno, T., Suzuki, T., Minami, N., and Imai, H. (2005). Stage-specific expression of microRNAs during Xenopus development. *FEBS Lett.* **579,** 318–324.

Watanabe, T., Takeda, A., Tsukiyama, T., Mise, K., Okuno, T., Sasaki, H., Minami, N., and Imai, H. (2006). Identification and characterization of two novel classes of small RNAs in the mouse germline: Retrotransposon-derived siRNAs in oocytes and germ-line small RNAs in testes. *Genes Dev.* **20,** 1732–1743.

Yekta, S., Shih, I. H., and Bartel, D. P. (2004). MicroRNA-directed cleavage of HOXB8 mRNA. *Science* **304,** 594–596.

Yigit, E., Batista, P. J., Bei, Y., Pang, K. M., Chen, C. C., Tolia, N. H., Joshua-Tor, L., Mitani, S., Simard, M. J., and Mello, C. C. (2006). Analysis of the *C. elegans* Argonaute family reveals that distinct Argonautes act sequentially during RNAi. *Cell* **127,** 747–757.

CHAPTER TEN

Dissecting MicroRNA-Mediated Gene Regulation and Function in T-Cell Development

Tin Ky Mao *and* Chang-Zheng Chen

Contents

Abstract

MicroRNAs (miRNAs) are abundant ~22-nucleotide regulatory RNAs encoded in animal genomes. They are thought to exhibit diverse biological functions in animals by targeting messenger RNAs (mRNAs) for degradation or translational repression. Here we use T-cell development as a model to illustrate methods and strategies for dissecting the post transcriptional gene regulatory networks controlled by miRNAs and their roles in the differentiation of T-cell precursors. The process involves the identification of miRNA genes in rare T-cell progenitors, determining miRNA expression during T-cell development, characterizing miRNA function in T-cell development using an *in vitro* assay, and identifying functionally relevant gene(s) regulated by miRNAs.

Department of Microbiology and Immunology, Baxter Laboratory of Genetic Pharmacology, Stanford University School of Medicine, Stanford, California

Methods in Enzymology, Volume 427
ISSN 0076-6879, DOI: 10.1016/S0076-6879(07)27010-7

1. INTRODUCTION

Discoveries of RNA interference (RNAi) have revealed a fundamental layer of gene regulation controlled by small noncoding RNAs (Elbashir *et al.*, 2001; Fire *et al.*, 1998; Lee *et al.*, 1993; Wightman *et al.*, 1993). Among these small noncoding RNAs, miRNAs are a family of endogenous small RNAs that mediate posttranscriptional gene repression through imperfect base-pairing to their cognate target mRNAs (Bartel, 2004). miRNA-mediated gene regulation is likely to represent a fundamental layer of genetic programs at the posttranscriptional level. miRNA genes are abundant in animals and account for over 5% of human protein-coding genes (Berezikov *et al.*, 2005; Miranda *et al.*, 2006). Furthermore, each miRNA can regulate a large number of protein-coding genes (Lewis *et al.*, 2003, 2005; Lim *et al.*, 2005; Miranda *et al.*, 2006). Supporting the notion that miRNA may have broad impacts on gene expression (Bartel, 2004), miRNAs are found to play diverse functions in animals (Wienholds and Plasterk, 2005), including timing of development and neuronal patterning in worms (Lee *et al.*, 1993; Wightman *et al.*, 1993), regulation of cell proliferation, cell death, fat metabolism in flies (Brennecke *et al.*, 2003; Xu *et al.*, 2003), modulating hematopoietic lineage differentiation (Chen *et al.*, 2004), granulopoiesis (Fazi *et al.*, 2005), insulin secretion (Poy *et al.*, 2004), adipocyte differentiation (Esau *et al.*, 2004), and heart development (Zhao *et al.*, 2005) in vertebrate animals. These reports represented the first few lines of evidence exposing the many functions of miRNAs in vertebrate animals, lending further support that miRNAs are essential components of the molecular circuitry that controls many, if not all, biological processes (Bartel and Chen, 2004).

In this chapter, we present a systematic approach to dissect the role of miRNA-mediated gene regulation and function in T-cell development (outlined in Fig. 10.1). We will focus on discussing methods that can be applied to characterize miRNA expression in rare progenitor cells, to interrogate miRNA function in T-cell development, and to identify functional target genes. Strategies and methods discussed here provide an example for dissecting miRNA-mediated gene regulation and function in T-cell development and should be broadly applicable to study miRNA function in other progenitor cells.

2. CHARACTERIZING miRNA EXPRESSION DURING T-CELL DEVELOPMENT

The initial step in elucidating the role of miRNAs in T-cell development is to characterize miRNA expression in developing thymocyte populations (see Fig. 10.1A). The differentiation of immature T cells in the

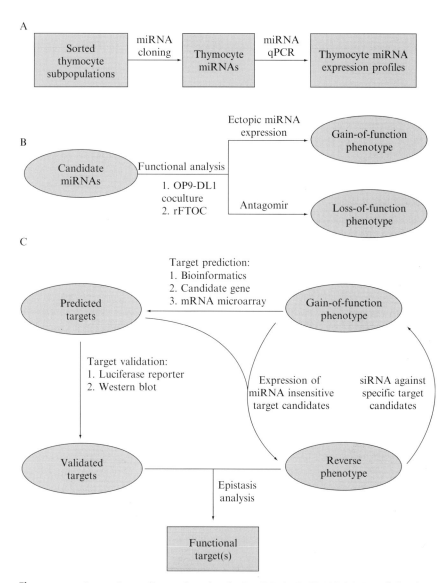

Figure 10.1   Strategies to dissect the role of microRNAs (miRNAs) in T-cell development. (A) Identification of miRNAs expressed in thymocytes, including rare subpopulations, is made possible using a sensitive miRNA cloning protocol. Quantitative PCR are subsequently used to validate cloning analysis and to provide an miRNA expression profile unique for each population. (B) To characterize miRNA function during T-cell differentiation, miRNA can be ectopically expressed or silenced by antagomirs to provide gain–of-function and loss–of-function phenotypes, respectively. OP9-DL1 coculture assay and rFTOC are two *in vitro* systems in which T-cell differentiation can be monitored by modulating the level of the miRNA. (C) An approach to identify and

thymus involves sequential steps characterized by critical molecular and biological events, such as T-cell receptor (TCR) recombination, and positive and negative selection. Each maturation step is defined by a distinct combination of cell surface markers, which permits the isolation of all thymic subsets by flow cytometry. Early thymic progenitors are double-negative (DN) for CD4 and CD8, which can be further fractionated based on their unique expression of CD25 and CD44 in the following order: DN1 ($CD44^+$ $CD25^-$), DN2 ($CD44^+$ $CD25^+$), DN3 ($CD44^-$ $CD25^+$), and DN4 ($CD44^-$ $CD25^-$). Upon the complete rearrangement of TCR α and β chains, the immature thymocytes enter a CD4 and CD8 double-positive (DP) stage. It is at the DP stage when T-cell clones bearing immunocompetent TCRs are then selected to become mature CD4 or CD8 single-positive (SP) cells. The ability to isolate various T-cell populations in thymus sets a foundation for the identification and characterization of miRNA expression during T-cell development.

The rare nature of certain thymic populations, such as DN1 to DN4, poses a unique challenge for miRNA expression profiling. Many existing approaches for quantifying miRNA expression are limited in sensitivity or specificity for these rare thymic progenitor cells. More importantly, it has been estimated that a large number of miRNA genes remains to be identified (Berezikov et al., 2005; Miranda et al., 2006). Furthermore, miRNAs expressed in rare progenitor cells are likely missing from the public databases, which contain miRNAs mainly cloned from cell lines and tissues. Therefore, the use miRNA microarray for global analyses of miRNA expression in rare progenitor cells is biased toward known miRNAs, and progenitor-specific miRNAs may be missing from the chip.

2.1. miRNA cloning analysis

To systematically characterize miRNA expression during T-cell development, one can first construct miRNA libraries from rare progenitor cells using a sensitive miRNA cloning protocol (Min and Chen, 2006) and then sequence the miRNA libraries to determine the identities and the frequencies

---

validate functionally relevant target genes. Potential targets are predicted through computational algorithms, candidate gene approach, or mRNA microarray, and validated by luciferase reporter assays and Western blots. Expression of candidate targets with predicted binding sites either deleted or mutated (miRNA-insensitive) should reverse the miRNA gain-of-function phenotype if the target genes are functionally relevant. In contrast, specific silencing target candidates by shRNA should recapitulate the miRNA overexpression phenotype if the repression of a single target by the miRNA is sufficient for its function. Epistatic analysis is used to validate functionally relevant target by manipulating the miRNA-target gene interaction and correlating with functional phenotype.

of miRNAs cloned. MiRNA cloning frequencies can be used as an indicator for relative abundance of an miRNA in the cell types surveyed. Under the assumption that cell size and total miRNA abundance are similar between the cell types compared, cloning frequencies may be directly used to evaluate miRNA abundance between different samples. Sequencing a large number of miRNAs from each miRNA library should make the cloning frequency statistically more robust and quantitative. Because the cell size and total miRNA abundance varies significantly during T-cell development in thymus, one may have to consider the aforementioned factors when comparing miRNA expression levels between different thymic progenitor cells when using miRNA cloning frequencies as indicators for abundance.

2.2. miRNA qPCR analysis

Although miRNA cloning allows for the identification of known and novel miRNAs expressed in thymic progenitor cells, quantitative miRNA PCR can serve to validate cloning frequencies as well as to precisely determine miRNA expression level in various progenitor cell populations. The ability to quantitate miRNAs in rare cell populations by PCR approach has been hampered by specificity and sensitivity issues. However, one method seemed to be able to circumvent these obstacles by the addition of a stem-loop reverse transcriptase (RT) primer specific for each miRNA followed by traditional TaqMan PCR analysis (Chen *et al.*, 2005). As a result, only 5 ng of total RNA is required for the detection of a single miRNA. Taq-Man primer and probe sets for many known miRNAs are commercially available through Applied Biosystems. (Please refer to manufacturer instructions for detailed protocols.) Because one can infer which miRNAs are absent or expressed at very low levels in thymic progenitor cells based on cloning analyses, fewer qPCR assays need to be carried out to compare miRNA expression between different thymic progenitors. miRNA qPCR analyses can be used for relative and absolute miRNA quantification in thymic progenitor cells.

   To determine the exact copy number of an miRNA in a rare progenitor cell type, one can carry out absolute quantitation with miRNA qPCR assay. By spiking the purified thymic progenitor cells with a precise amount of synthetic miRNA (i.e., 50,000 copies per cell), a ratio of miRNA of interest to the spiked miRNA is established and is assumed to remain constant through RNA isolation, RT, and qPCR analyses. Then, exact copies of test and spiked miRNA in defined amount of total RNA input (typically 5 ng per qRT/PCR reaction) are determined by using standard curves for the test and spiked miRNA. Once the quantities are determined for the test and spiked miRNA, the measured ratio can be used to determine the number of copies of the test miRNA originally present in each cell type according to the initial amount of spiked miRNA. The following is a detailed protocol for determining the exact copies of an miRNA in rare progenitor cells using TaqMan miRNA assays:

1. A defined number of thymic progenitor cells are sorted by FACS and immediately lysed in TriZol Reagent. Prior to total RNA isolation, spike a known amount of cells with a synthetic miRNA. It is important that your choice of spiked miRNA is absent or expressed at very low levels in the cells of interest. We have determined that miR-122a, a liver-specific miRNA, is a valid spike for thymocyte populations. We generally spike 50,000 copies of synthetic miRNA per cell and then isolate total RNA using TriZol Reagent according to manufacturer's protocol. To assist in the visualization of the RNA pellet, we add glycogen (1 μl of 20 mg/ml stock) as a carrier during RNA precipitation.

2. Carry out RT reaction utilizing stem-loop RT primers against the test and spiked miRNAs (Chen et al., 2005). The RT reaction is performed on test samples as well as synthetic miRNAs to generate a standard curve. We normally generate a 5-point standard curve based on tenfold serial dilutions ranging from 0.01 to 100 pM. To ensure similar RT efficiencies between reactions containing test samples or synthetic RNA, 5 ng of bacterial RNA is introduced as carrier along with reactions containing only synthetic miRNA. Triplicate RT reactions are performed for each sample and each standard curve point. See Table 10.1 for a list of components for RT reaction. The PCR thermocycler conditions are: (1) 16°, 30 min; (2) 42°, 30 min; and (3) 85°, 5 min.

3. One microliter of the RT reaction is used for subsequent real-time PCR reactions. See Table 10.1 for a list of components for TaqMan qPCR reaction. The real-time PCR thermocycler conditions are: (1) enzyme activation at 95°, 10 min; (2) denature at 95°, 15 sec; (3) anneal/extend at 60°, 60 sec; and (4) repeat steps 2 and 3 for a total of 40 cycles.

4. The quantities (in picomolar) of the miRNA of interest and spiked miRNA can be extrapolated from the standard curves. The number of copies of miRNA per cell can be determined based on the assumption that the ratio of miRNA of interest to spiked miRNA remains constant:

$$\frac{\text{miRNA of Interest (p}M)}{\text{spiked miRNA (p}M)} = \frac{X}{50,000 \text{ copies/cell of spiked miRNA}}$$

$$X = \text{miRNA of Interest (copies/cell)}$$

$$X = (\text{miRNA:spike ratio}) \, (50,000 \text{ copies/cell})$$

$$\text{Eq. (1)}$$

Table 10.1 Components of reverse transcriptase and TaqMan qPCR reactions for miRNA expression analysis

Reverse transcriptase mix	vol/7.5 μl rxn	qPCR mix	vol /10 μl rxn
10 mM dNTPs[a]	0.75 μl	DEPC-treated H$_2$O	3 μl
10× RT Buffer[a]	0.75 μl	TaqMan 2× PCR Master Mix[d]	5 μl
Multiscribe^{TM} RT (50 U/μl)[a]	0.5 μl	10× TaqMan Probe/ Primers Mix	1 μl
RNase Inhibitor (20 U/μl)[a]	0.1 μl	cDNA	1 μl
5× RT Primer[b]	1.5 μl		
DEPC-treated H$_2$O	2.9 μl		
RNA Sample[c]	1.0 μl		

[a] These reagents are provided in the High Capacity cDNA Reverse Transcriptase Kit (Applied Biosystems).
[b] RT primers and TaqMan probe/primers mix are provided in TaqMan microRNA assays (Applied Biosystems) in 5× and 20× format, respectively.
[c] For reactions containing only synthetic miRNA, add 5 ng of carrier bacterial RNA (Ambion).
[d] TaqMan 2× Universal PCR Master Mix (Applied Biosystems) contains DNA polymerase, dNTPs, and optimized buffer components.

Please note that if no spike is used, one can utilized the amount of total RNA input (5 ng) to determine the amount of miRNA per mass of total RNA (copies per pg):

$$\frac{\text{miRNA of interest}(pM)* \text{RT vol} *\text{Avogadro's no.}}{\text{input RNA}(pg)} \qquad \text{Eq. (2)}$$

As an alternative to absolute quantification, one can use comparative Ct (critical threshold) method for relative quantification. Relative changes of miRNA expression are determined by comparing the miRNA expression in various cell populations to a reference sample, also known as a calibrator. This method requires the normalization of miRNA detection to an endogenous control gene by calculating the ΔCt (test miRNA Ct minus control gene Ct) for each experimental and calibrator sample. Changes in miRNA expression can be expressed as changes in the ΔCt values (ΔΔCt) between an experimental and calibrator sample or relative fold difference ($2^{-\Delta\Delta Ct}$). Small nuclear RNAs (snoRNAs) are representative housekeeping genes for normalizing miRNA expression because they are abundantly expressed in many cell types and their expression levels in different cell types are thought to remain constant. Furthermore, qPCR assays that measure snoRNAs are based on the same chemistry as that used for miRNA qPCR. One can also compare the relative abundance of different miRNAs using this method, assuming that probes set for different miRNAs have the same amplification efficiency. By combining absolute and relative quantitation, one can determine miRNA expression levels in various thymic progenitor cells, thus suggesting possible candidate miRNAs that may play important roles in T-cell development and function in thymus.

3. RETROVIRAL CONSTRUCTS FOR miRNA EXPRESSION

Gain-of-function analyses may be used to perturb candidate miRNA expression in thymic progenitors and to reveal their potential biological roles in the T cell developmental processes. We have developed retroviral vectors for stable expression of miRNAs in hematopoietic stem/progenitor cells (Chen *et al.*, 2004). The original vector was constructed using murine stem cell retrovirus backbone with the miRNA expression cassette driven by an RNA polymerase III (human H1 promotor) placed in the U3 region of the 3′-long terminal repeat (LTR) (Fig. 10.2A). We have also shown that corresponding genomic sequences that flank the pre-miRNAs are required for efficient miRNA processing and we typically insert miRNA genes with 125-nt flanking sequences into the H1 expression cassette (Chen *et al.*, 2004). Although some miRNAs are thought to be endogenously transcribed by the pol II polymerase (Lee *et al.*, 2004), we have found that miRNA genes expressed from pol II or pol III polymerases in our retroviral vectors are functionally indistinguishable (Mao, T. K. and Chen, C.-Z., unpublished observation).

Nevertheless, this pol III promoter-based double-copy retroviral miRNA expression vector has some limitations. First, the pol III H1 promoter is known to be inefficient in making large transcripts. Second, some miRNA genes may contain a stretch of consecutive A's in their flanking sequences that can serve as termination signals for pol III H1 transcription. Finally, large inserts dramatically reduce the viral packaging efficiency because the H1 expression cassette is localized in the 3′ LTR. It is known that the length of certain pri-miRNAs can extend beyond 1 kb, containing multiple miRNAs and exceeding the length of typical pol III transcripts (Bartel, 2004). To overcome these issues we have modified the retroviral vector to express miRNAs from the pol II promoters.

In our MXW-PGK-IRES-GFP vector, the bicistronic pol II cassette contains an internal ribosome entry site (IRES) flanked by the miRNA hairpin upstream and green fluorescent protein (GFP) downstream (Fig. 10.2B). The coexpression of the miRNA and GFP is controlled by the constitutive murine 3-phosphoglycerate kinase promotor (P_{PGK}), which allows for

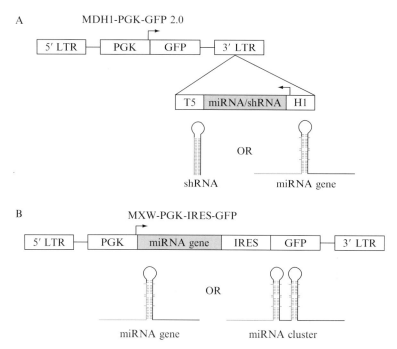

Figure 10.2 MicroRNA (miRNA) retroviral expression vectors. (A) Expression of miRNA hairpins or short hairpin RNAs (shRNAs) by RNA polymerase III, a human H1 promotor, in the MDH1-PGK-GFP2.0 vector. The pol III expression cassette is configured in reverse orientation within the 3′ LTR to confer double-copy expression. For miRNA expression, an miRNA gene with approximately 125 nt of its own genomic flanking sequences inserted into the H1 expression cassette. (B) Expression of miRNA hairpins or clusters by RNA polymerase II, murine PGK promotor, in the MXW-PGK-IRES-GFP vector. The pol II bicistronic expression cassette allows for the constitutive coexpression miRNA(s) and GFP. The expression of miRNAs from pol II requires an miRNA gene with ∼250 nt of its own genomic flanking sequences placed downstream to PGK promotor.

virus-infected cells expressing the miRNA to be monitored by GFP fluorescence using a flow cytometer. We have used this expression construct to successfully generate high-titer virus containing various miRNAs that consistently led to high infection of thymic progenitor cells. The use of pol II promoters, such as P_{PGK}, has been demonstrated to facilitate the transcription of the longer pri–miRNAs (Cullen, 2005; Dickins *et al.*, 2005; Silva *et al.*, 2005). We have found that the use of pol II is particularly effective when expressing two distinct mature miRNAs that are separated up to, but not limited to, 1 kb due to the strength of the PGK promotor. Indeed, analysis of miRNA gene clustering patterns revealed that they are evolutionarily conserved and constitute 37% of human miRNAs, signifying that clustering of miRNA genes possesses functional implications (Altuvia *et al.*, 2005). The MXW–PGK–IRES–GFP vector, therefore, provides an avenue by which to express miRNA clusters and examine their functional significance in T–cell development. We generally insert miRNA genes with at least 250 base pairs flanking both sides of the miRNA (~520 bp in total) into the expression cassette. Using miRNA genes shorter than ~520 base pairs may lead to lower expression levels. Detailed protocols for vector construction, viral packaging, and infection have been previously described (Min and Chen, 2006).

4. INVESTIGATING miRNA FUNCTION IN T-CELL DEVELOPMENT

The progressive maturation of T cells within the thymus is perhaps one of the most investigated developmental processes in mammals. However, miRNAs add yet another dimension by which this intricate biological process is regulated. For many years, fetal thymic organ cultures (FTOC) have been regarded as the only practical *in vitro* system to recapitulate the normal thymic microarchitecture that support the proper differentiation of T cells (Hare *et al.*, 1999). However, this experimental system was rather inefficient to initiate and facilitated only limited cellular yield, thereby limiting their practical application for large-scale analyses. To examine the functions of miRNA in T-cell development, we have adapted a simple and efficient *in vitro* assay initially developed by the Zuniga-Pflucker lab based on the coculture of thymic progenitors on a monolayer of OP9 stromal cells expressing the Notch ligand Delta-1 (Balciunaite *et al.*, 2005; de Pooter and Zuniga-Pflucker, 2007; Schmitt and Zuniga-Pflucker, 2002). In the OP9-DL1 cultures, murine fetal liver stem cells were shown to recapitulate the normal T-cell developmental program, including undergoing marked clonal expansion and giving rise to both $\alpha\beta$- and $\gamma\delta$-T cells (Schmitt and Zuniga-Pflucker, 2002). By establishing a means to ectopically express miRNAs in primary thymocytes, we are able to address the biological roles of miRNAs in T-cell development using the OP9-DL1 coculture assay. Here is a detailed protocol on how to examine miRNA function in T-cell development using OP9-DL1 stromal coculture assay (Fig. 10.3).

(1) Prepare thymocytes for retroviral infection:

Four days prior (Day −4) to coculture, 6–week-old C57BL/6 mice were intravenously administered $300\ \mu l$ 5-fluorouracil (5-FU) via the retro-orbital or lateral tail vein at a dose of 150 mg/kg body weight. *In vivo* treatment of 5-FU serves two purposes: (1) causing cell death of cycling thymocytes—we observed DN cells are enriched as a result of significant cell loss stemming from the DP compartment; and (2) stimulating quiescent thymic progenitors into cell cycle which facilitates retroviral infection.

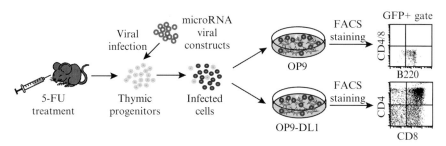

Figure 10.3 Schematic diagram of the OP9-DL1 stromal coculture assay to monitor T-cell development. Mice are primed with 5-FU 4 days prior to retroviral infection of the miRNA expression constructs. Infected T-cell progenitors are cultured over a monolayer of OP9-DL1 stromal cells for 8 days. FACS analysis is used to determine the effect of ectopic miRNA expression on absolute percentages thymocyte populations by gating on miRNA expressing thymocytes (CD45$^+$GFP$^+$).

Thymocytes are harvested following 4 days of 5-FU priming (Day 0). Typically, we are able to extract 3 to 5 × 10^6 thymocytes per mouse treated with 5-FU.

(2) Prepare OP-DL1 stromal cells:

On Day −2, OP9-DL1 stromal cells are trypsinized and seeded in 24-well plates at 20,000 cells per well. At this cell density, the stromal cells should be confluent on Day 0, the time of coculture initiation.

(3) Retroviral infection of thymocytes and culture over OP9-DL1:

Following isolation of 5-FU-primed thymocytes, they are infected by retroviral supernatant via spinoculation. Because we generally perform 12 replicates per miRNA-expression construct and seed 100,000 thymocytes per well, a total of 1.2 × 10^6 thymocytes are needed per infection. The thymocytes are resuspended in 1 ml of OP9-DL1 medium and mixed with equal vol of viral supernatant (at least 10^6 cfu/ml). If viral titer is below optimal, you may increase the vol of viral supernatant. To assist in retroviral infection, polybrene is supplemented in virus/thymocyte mix at a final concentration of 4 μg/ml. The spinoculation is performed at 2000 rpm for 2 h at room temperature. Upon completion of spin infection, carefully aspirate off the supernatant and resuspend cells in T-cell culture medium to a density of 0.2 × 10^6 per ml. The culture media are removed from the OP9-DL1 monolayer and 500 μl of cell suspension is seeded onto the stromal layer to initiate the coculture (Day 0).

On day 1, nonadherent thymocytes are suspended in the culture by gentle tapping of the plates. This is quickly followed by aspiration of the culture medium, and 500 μl of fresh T-cell culture medium is used to replenish the coculture. We have determined that, by sorting individual thymocyte population and plating each population onto OP9-DL1 stromal layer, the adherent cells composed of DN1–3, whereas DN4, DP, and SP thymocytes are nonadherent and are removed by this washing step. Hence, any development of DP thymocytes on the day of termination are likely due to the contribution of adherent DN1–3 cells. On day 6 of the coculture, an additional 500 μl of fresh T-cell culture medium is added to each well. We observed that during the next 2 days until termination, the thymocytes within the culture undergo significant expansion.

(4) Termination and Preparation for FACS Analysis:

The coculture is terminated on Day 8 and analyzed by FACS to determine the fraction of DP cells in the culture. The first step in the preparation of cells for FACS analysis is to remove nonadherent cells from the culture medium into 1.2-ml cluster tubes. These cluster tubes are preferred over conventional FACS tubes because their 96-well format facilitates preparation of many samples. The cells are pelleted by centrifugation and the supernatant are removed using an 8-channel aspirator. To assist in the dissociation of adherent thymocytes from the OP9-DL1 stromal cells, we add 100 μl of 0.8 mg/ml collagenase type IV to the monolayer and incubate at 37° for 15 min. Because collagenase type IV contains a small percentage of trypsin, extensive exposure to collagenase will damage the integrity of surface molecules and compromise their ability to be recognized by monoclonal antibodies in subsequent FACS analysis. Thus, it is critical that the enzymatic reaction be terminated immediately with 500 μl ice-cold PBS. It is prudent for an investigator to determine the maximal collagenase exposure time without the loss of detection for each antibody used for flow cytometry. Following treatment with collagenase, the monolayer can easily be dissociated by gentle repetitive pipetting and transferred to cluster tubes containing the nonadherent cells.

The mixture of adherent and nonadherent cells are filtered through a 50-μm nylon mesh and washed twice with ice-cold FACS buffer. Upon the final wash, the cell pellet is resuspended in 100 μl of FACS buffer and stained with antibodies against CD4 and CD8. It is important to note that, in addition to infected thymocytes, the OP9-DL1 cell line also expresses GFP. The majority of OP9-DL1 cells can be separated from T cells based on the size gate (FSC vs. SSC). However, when the infection rate is low or there is insufficient expansion of thymocytes in the culture, OP9-DL1 contamination can significantly lower the percentages of CD4$^+$CD8$^+$ DP thymocytes by increasing the percentage of DN cells. Therefore, we use CD45 staining in FACS analysis to differentiate thymocytes from stromal cells.

To analyze the effect of miRNA overexpression on the absolute percentages of DP thymocytes, the samples are gated for their positive expressions of CD45 and GFP to mark infected thymocytes. Although we have utilized the percentages of DP thymocytes as an indicator for functional output, it is also important to note that DN thymocytes are multipotent and can differentiate into other lineages, including $\gamma\delta$-T cells or natural killer

(NK) cells. Therefore, a decrease percentage of DP cells with a concomitant increase in DN cells could signify a possible diversion into another lineage caused by ectopic miRNA expression.

The OP9-DL1 assay offers an efficient and robust means to address functional roles of miRNAs in T-cell development. However, this system is not without limitations. For example, OP9-DL1 stromal cells are unable to support the differentiation of $CD4^+$ T cells and NKT cells due to the lack of MHC class II molecules and CD1d expression, respectively (Zuniga-Pflucker, 2004). Another drawback is that OP9-DL1 cells originated from a bone marrow stromal cell line, which may be intrinsically different in their ability to present a comprehensive repertoire of self-antigens to developing T cells (Zuniga-Pflucker, 2004). As a result, this system may not be appropriate to address fundamental questions involving positive and negative selection. Reaggregate FTOC (rFTOC) represent an alternative assay to monitor miRNA function (see Fig. 10.1B).

5. IDENTIFICATION AND VALIDATION OF FUNCTIONALLY RELEVANT miRNA TARGET GENES

To understand the molecular mechanisms by which miRNAs control T-cell development, it is essential to identify the genes regulated by miRNAs. A number of computational algorithms have been developed to predict potential genes targeted by miRNAs in vertebrates (Kiriakidou *et al.*, 2004; Krek *et al.*, 2005; Lewis *et al.*, 2003, 2005; Miranda *et al.*, 2006). Although there are inherent differences among these prediction programs, they generally require that target sites form perfect base pairing two to seven nucleotides from the 5' end of the miRNA, known as the "seed" region, and that the seed pairing is conserved across species (Rajewsky, 2006). There are currently a plethora of online resources for miRNA target prediction (refer to Rajewsky [2006] for a comprehensive list). However, it remains a challenge to select the functionally relevant targets from the long list of predicted ones.

A few approaches may be adopted to narrow down the targets from the long list of predicted ones. Studies have demonstrated that at least some miRNAs can induce degradation of mRNAs (Bagga *et al.*, 2005; Lim *et al.*, 2005). Thus, a complement to computational predictions is to directly correlate miRNA overexpression with the change of mRNA levels. For example, mRNA expression microarray analyses may be used to reveal the effects of ectopic miRNA expression on the mRNA levels. One can then select for genes whose transcripts are significantly down regulated and concomitantly possess potential miRNA binding sites. An alternative approach is to take advantage of known molecular insights of a particular biological phenotype and associate that knowledge with the miRNA gain-of-function phenotypes. In the case of T-cell development, established molecular and cellular events that govern T-cell development allows one to connect proteins known to regulate T-cell differentiation with an observed miRNA gain-of-function phenotype. By incorporating a candidate gene approach based on information extrapolated from miRNA function, one can significantly narrow the number of candidate target genes from a long list of computational predicted targets. However, it should be noted that target genes regulated at the translational level will be missed in the mRNA microarray analyses, and target genes whose function are still unknown will likely to be left out by the candidate gene approach.

Once the candidate target genes are identified, various *in vitro* cell culture analyses, such as luciferase reporter and Western blot assays, may be employed to validate the direct regulation of predicted target genes by a miRNA (Lewis *et al.*, 2003; Lim *et al.*, 2005; Miranda *et al.*, 2006). However, although the cell culture-based validations provide evidence for the potential physical interaction between a miRNA and their cognate target genes, it does not establish functional interaction between the targets and the miRNA in T-cell development. Therefore, with the validation of multiple target genes by luciferase reporter and Western blot assays, it is then critical to determine whether the regulation of one or many of the validated target gene(s) is important for the function of the miRNA in T-cell development. Genetic approaches using mice to establish the functional interactions between a miRNA and its cognate targets could be a daunting task. The advantage of the *in vitro* OP9-DL1 coculture assay is that this functional assay allows one to manipulate epistatic interactions between a miRNA and its target(s) using the T-cell development as a functional read out.

To determine whether the down regulation of one or more of the confirmed target gene(s) is required for miRNA function, one can use short hairpin RNAs (shRNAs) designed to target individual genes and examine the consequence of silencing this target gene on thymic progenitor cell differentiation using OP9-DL1 coculture assay. The general strategies and factors to consider for designing shRNAs are presented in a review by the Alberola-Ila lab (Hernandez-Hoyos and Alberola-Ila, 2005). The pol III promoter-based retroviral vector (MDH1-PGK-GFP2.0, Fig. 10.2A) can be used to express shRNAs in thymic progenitor cells (Li *et al.*, 2007). If regulation of a single target gene by the miRNA is sufficient for its function, then down regulation of this target gene expression by specific shRNAs should recapitulate the phenotype of miRNA ectopic expression in thymic progenitor cells.

However, if down regulation of a single target gene by specific shRNAs cannot (or only partially) recapitulate the phenotype of miRNA ectopic expression in thymic progenitor cells, then there are two possible explanations: one is that all targets are required for miRNA function, and the other is that none of the targets are required for miRNA function. Because miRNAs mediate their effects by repressing gene expression at post-transcriptional levels, restoring the expression of a functional target gene, which is normally repressed by the endogenous or ectopically expressed miRNAs, will result in a phenotype that opposes the miRNA overexpression phenotype. Therefore, one can ectopically express individual candidate genes that lack sites for miRNA binding in thymic progenitor cells and examine their function in T-cell development in the OP9-DL1 coculture assay. Specifically, only the coding region or the full-length cDNA of the target gene with potential miRNA-binding sites mutated can be cloned downstream to the PGK promoter in the MXW-PGK-IRES-GFP vector (Fig. 10.2B). If overexpression of a miRNA-insensitive target produces a phenotype that opposes miRNA ectopic expression, then this target is likely a functionally relevant target, whereas the absence of a reverse phenotype is likely to signify an irrelevant target. A more convincing approach would be to determine whether restoration of target expression can reverse the miRNA overexpression phenotype by coexpressing the miRNA and the miRNA-insensitive target in thymic progenitors using the MDH1-PGK-IRES-GFP vector (Fig. 10.4A; vector construction described later).

Overexpression of candidate target genes that display the opposite phenotype as ectopic miRNA expression can be validated further as a functional target through an epistatic analysis to modify the interaction between two genes: the miRNA and its target (Fig. 10.4). This would require simultaneous expression of both the miRNA and its target in thymic progenitors. To facilitate the coexpression, PGK and H1 promotors are used to drive the expression of the full-length target cDNA (ORF + 3' UTR) and the miRNA, respectively in the MDH1-PGK-IRES-GFP vector (see Fig. 10.4A). By coexpressing wild-type (wt) miRNA and wt target (see Fig. 10.4B, i), the target gene regulation by the miRNA is intact and should lead to an observed biological function similar to the miRNA gain-of-function phenotype. It has been previously shown through systematic target-site mutation experiments that target sites containing sufficient complementarity to the miRNA 5' end are critical for biological function (Brennecke et al., 2005). Therefore, by generating mutations in either the miRNA 5' seed region (see Fig. 10.4B ii) or the target sites that bind to the seed region (see Fig. 10.4B, iii), one can test whether disrupting the epistatic interaction can abolish functional output. The validation of the target as functionally relevant is evident if the activity can be restored through the generation of compensatory mutations in the miRNA seed region that complements the mutated target binding sites (see Fig. 10.4B, iv). One caveat to keep in mind when restoring the miRNA target interaction is that the functional activity may only be partially restored due to the existence of cryptic target sites. Moreover, the number of target sites necessary for

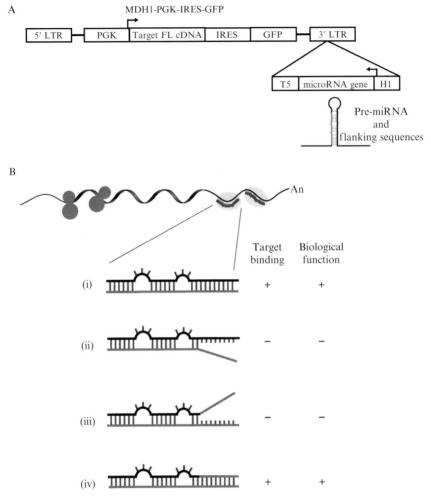

Figure 10.4 (A) To facilitate epistatic analysis, the MDH1-PGK-IRES-GFP vector allows for simultaneous expression of the target gene and the microRNA (miRNA). This vector is composed of two expression cassettes: the PGK promoter drives the expression of the target full-length cDNA and the H1 promotor drives the expression of the miRNA hairpin. (B) Schematic diagrams describing the possible epistatic interactions and predicted biological outcomes between the target and the miRNA: (i) wt/wt, (ii) wt/mut, (iii) mut/wt, (iv) mut/mut interactions.

regulation by a miRNA may differ for each target gene and must be determined empirically. Cumulatively, the process of target prediction, validation, and identification of functionally relevant targets will allow one to thoroughly dissect the molecular mechanisms through which candidate miRNAs exert their control of T-cell development.

6. MATERIALS AND REAGENTS

RT-qPCR

TriZol Reagent (Invitrogen, Cat. No. 15596–018)
Chloroform
100% Isopropyl Alcohol
75% and 100% Ethanol
DEPC-treated H_2O (Ambion, Cat. No. 9915G)
Glycogen (Roche Applied Science; Cat. No. 10901393001)
High-capacity cDNA Reverse Transcriptase Kit (Applied Biosystems; Cat. No. 4368813)
Taqman® MicroRNA Assays (Applied Biosystems; unique Cat. No. for each miRNA assay)
Taqman® 2× Universal PCR Master Mix (Applied Biosystems; Cat. No. 4304437)
E. coli Total RNA (Ambion; Cat. No. 7940)

OP9-DL1 Coculture Assay

5-Fluorouracil (Sigma–Aldrich; Cat. No. F6627)
C57BL/6 mice (Jackson Laboratory)
0.05% Trypsin–EDTA (Invitrogen; Cat. No. 15400)
Polybrene (Hexadimethrine Bromide; Sigma-Aldrich; H9268)
OP9-DL1 Culture Medium: Minimum Essential Medium-α (α-MEM; Invitrogen; Cat. No. 32561), 20% Fetal Bovine Serum (FBS; HyClone; Cat. No. SV30014.03), and Penicillin-Streptomycin (Invitrogen; Cat. No. 15140).
T cell Culture Medium: α-MEM; 20% Characterized FBS (Hyclone; Cat. No. SH30071.03), Pen-Strep, 5×10^{-5} M 2-Mercaptoethanol (Invitrogen; Cat. No. 21985-023), 10 mM HEPES Buffer Solution (Invitrogen; Cat. No. 15630-080), 1 mM Sodium Pyruvate (Invitrogen; Cat. No. 11360-070), 5 ng/ml IL-7 (PeproTech; Cat. No. 217-17), and 27.5 ng/ml Flt3-Ligand (PeproTech; Cat. No. 300-19)
Cluster Tubes; 96-well format (Costar; Cat. No. 4410)
Collagenase Type IV (Worthington; Cat. No. LS004188): Reconstitute in Hank's Balanced Salt Solution with calcium (HBSS; Invitrogen; Cat. No. 14025)
FACS buffer: Dulbecco's Phosphate Buffered Saline without calcium (Invitrogen; Cat. No. 14190), 2% FBS
50-μM Nylon Mesh (Small Parts, Inc.; Cat. No. CMN-0053)
Antibodies:
PE-α-mouse-CD45 (clone: 30-F11; BD Pharmingen; Cat. No. 553081)
PE-Cy5.5-α-mouse-CD4 (clone:L3T4; eBioscience; Cat. No. 35-0042-82)
APC-α-mouse-CD8a (clone: 53-6.7; eBioscience; Cat. No. 17-0081-83).

ACKNOWLEDGMENTS

We thank Gwen Liu, Joost Kluiver, other members of the Chen lab for the helpful discussions. This work was supported by NIH R01-HL081612 (to C.-Z. C.).

REFERENCES

Altuvia, Y., Landgraf, P., Lithwick, G., Elefant, N., Pfeffer, S., Aravin, A., Brownstein, M. J., Tuschl, T., and Margalit, H. (2005). Clustering and conservation patterns of human microRNAs. *Nucleic Acids Res.* **33**, 2697–2706.

Bagga, S., Bracht, J., Hunter, S., Massirer, K., Holtz, J., Eachus, R., and Pasquinelli, A. E. (2005). Regulation by *let-7* and *lin-4* miRNAs results in target mRNA degradation. *Cell* **122**, 553–563.

Balciunaite, G., Ceredig, R., Fehling, H. J., Zuniga-Pflucker, J. C., and Rolink, A. G. (2005). The role of Notch and IL-7 signaling in early thymocyte proliferation and differentiation. *Eur. J. Immunol.* **35**, 1292–1300.

Bartel, D. P. (2004). MicroRNAs: genomics, biogenesis, mechanism, and function. *Cell* **116**, 281–297.

Bartel, D. P., and Chen, C. Z. (2004). Micromanagers of gene expression: the potentially widespread influence of metazoan microRNAs. *Nat. Rev. Genet.* **5**, 396–400.

Berezikov, E., Guryev, V., van de Belt, J., Wienholds, E., Plasterk, R. H., and Cuppen, E. (2005). Phylogenetic shadowing and computational identification of human microRNA genes. *Cell* **120**, 21–24.

Brennecke, J., Hipfner, D. R., Stark, A., Russell, R. B., and Cohen, S. M. (2003). bantam encodes a developmentally regulated microRNA that controls cell proliferation and regulates the proapoptotic gene hid in Drosophila. *Cell* **113**, 25–36.

Brennecke, J., Stark, A., Russell, R. B., and Cohen, S. M. (2005). Principles of microRNA-target recognition. *PLoS Biol.* **3**, e85.

Chen, C., Ridzon, D. A., Broomer, A. J., Zhou, Z., Lee, D. H., Nguyen, J. T., Barbisin, M., Xu, N. L., Mahuvakar, V. R., Andersen, M. R., *et al.* (2005). Real-time quantification of microRNAs by stem-loop RT-PCR. *Nucleic Acids Res.* **33**, e179.

Chen, C. Z., Li, L., Lodish, H. F., and Bartel, D. P. (2004). MicroRNAs modulate hematopoietic lineage differentiation. *Science* **303**, 83–86.

Cullen, B. R. (2005). RNAi the natural way. *Nat. Genet.* **37**, 1163–1165.

de Pooter, R., and Zuniga-Pflucker, J. C. (2007). T-cell potential and development in vitro: the OP9-DL1 approach. *Curr. Opin. Immunol.* **19**, 163–168.

Dickins, R. A., Hemann, M. T., Zilfou, J. T., Simpson, D. R., Ibarra, I., Hannon, G. J., and Lowe, S. W. (2005). Probing tumor phenotypes using stable and regulated synthetic microRNA precursors. *Nat. Genet.* **37**, 1289–1295.

Elbashir, S. M., Leneckel, W., and Tuschl, T. (2001). RNA interference is mediated by 21- and 22-nucleotide RNAs. *Genes Dev.* **15**, 188–200.

Esau, C., Kang, X., Peralta, E., Hanson, E., Marcusson, E. G., Ravichandran, L. V., Sun, Y., Koo, S., Perera, R. J., Jain, R., *et al.* (2004). MicroRNA-143 regulates adipocyte differentiation. *J. Biol. Chem.* **279**, 52361–52365.

Fazi, F., Rosa, A., Fatica, A., Gelmetti, V., De Marchis, M. L., Nervi, C., and Bozzoni, I. (2005). A minicircuitry comprised of microRNA-223 and transcription factors NFI-A and C/EBPalpha regulates human granulopoiesis. *Cell* **123**, 819–831.

Fire, A., Xu, S., Montgomery, M. K., Kostas, S. A., Driver, S. E., and Mello, C. C. (1998). Potent and specific genetic interference by double-stranded RNA in *Caenorhabditis elegans*. *Nature* **391**, 806–811.

Hare, K. J., Jenkinson, E. J., and Anderson, G. (1999). *In vitro* models of T cell development. *Semin. Immunol.* **11**, 3–12.

Hernandez-Hoyos, G., and Alberola-Ila, J. (2005). Analysis of T-cell development by using short interfering RNA to knock down protein expression. *Methods Enzymol.* **392**, 199–217.

Kiriakidou, M., Nelson, P. T., Kouranov, A., Fitziev, P., Bouyioukos, C., Mourelatos, Z., and Hatzigeorgiou, A. (2004). A combined computational-experimental approach predicts human microRNA targets. *Genes Dev.* **18**, 1165–1178.

Krek, A., Grun, D., Poy, M. N., Wolf, R., Rosenberg, L., Epstein, E. J., MacMenamin, P., da Piedade, I., Gunsalus, K. C., Stoffel, M., *et al.* (2005). Combinatorial microRNA target predictions. *Nat. Genet.* **37**, 495–500.

Lee, R. C., Feinbaum, R. L., and Ambros, V. (1993). The *C. elegans* heterochronic gene *lin-4 encodes small RNAs with antisense complementarity to lin-14. Cell* **75**, 843–854.

Lee, Y., Kim, M., Han, J., Yeom, K. H., Lee, S., Baek, S. H., and Kim, V. N. (2004). MicroRNA genes are transcribed by RNA polymerase II. *EMBO J.* **23**, 4051–4060.

Lewis, B. P., Burge, C. B., and Bartel, D. P. (2005). Conserved seed pairing, often flanked by adenosines, indicates that thousands of human genes are microRNA targets. *Cell* **120**, 15–20.

Lewis, B. P., Shih, I. H., Jones-Rhoades, M. W., Bartel, D. P., and Burge, C. B. (2003). Prediction of mammalian microRNA targets. *Cell* **115**, 787–798.

Li, Q.-J., Chau, J., Ebert, P. J., Sylvester, G., Min, H., Liu, G., Braich, R., Manoharan, M., Soutschek, J., Skare, P., Klein, L. O., Davis, M. M., *et al.* (2007). miR-181a is an intrinsic modulator of T cell sensitivity and selection. *Cell* **129**, 147–61.

Lim, L. P., Lau, N. C., Garrett-Engele, P., Grimson, A., Schelter, J. M., Castle, J., Bartel, D. P., Linsley, P. S., and Johnson, J. M. (2005). Microarray analysis shows that some microRNAs downregulate large numbers of target mRNAs. *Nature* **433**, 769–773.

Min, H., and Chen, C. Z. (2006). Methods for analyzing microRNA expression and function during hematopoietic lineage differentiation. *Methods Mol. Biol.* **342**, 209–227.

Miranda, K. C., Huynh, T., Tay, Y., Ang, Y. S., Tam, W. L., Thomson, A. M., Lim, B., and Rigoutsos, I. (2006). A pattern-based method for the identification of MicroRNA binding sites and their corresponding heteroduplexes. *Cell* **126**, 1203–1217.

Poy, M. N., Eliasson, L., Krutzfeldt, J., Kuwajima, S., Ma, X., Macdonald, P. E., Pfeffer, S., Tuschl, T., Rajewsky, N., Rorsman, P., *et al.* (2004). A pancreatic islet-specific microRNA regulates insulin secretion. *Nature* **432**, 226–230.

Rajewsky, N. (2006). microRNA target predictions in animals. *Nat. Genet.* **38**(Suppl.), S8–13.

Schmitt, T. M., and Zuniga-Pflucker, J. C. (2002). Induction of T cell development from hematopoietic progenitor cells by delta-like-1 *in vitro. Immunity* **17**, 749–756.

Silva, J. M., Li, M. Z., Chang, K., Ge, W., Golding, M. C., Rickles, R. J., Siolas, D., Hu, G., Paddison, P. J., Schlabach, M. R., *et al.* (2005). Second-generation shRNA libraries covering the mouse and human genomes. *Nat. Genet.* **37**, 1281–1288.

Wienholds, E., and Plasterk, R. H. (2005). MicroRNA function in animal development. *FEBS Lett.* **579**, 5911.

Wightman, B., Ha, I., and Ruvkun, G. (1993). Posttranscriptional regulation of the heterochronic gene *lin-14* by *lin-4* mediates temporal pattern formation in *C. elegans. Cell* **75**, 855–862.

Xu, P., Vernooy, S. Y., Guo, M., and Hay, B. A. (2003). The Drosophila microRNA Mir-14 suppresses cell death and is required for normal fat metabolism. *Curr. Biol.* **13**, 790–795.

Zhao, Y., Samal, E., and Srivastava, D. (2005). Serum response factor regulates a muscle-specific microRNA that targets Hand2 during cardiogenesis. *Nature* **436**, 214–220.

Zuniga-Pflucker, J. C. (2004). T-cell development made simple. *Nat. Rev. Immunol.* **4**, 67–72.

MicroRNAs AND DISEASE

CHAPTER ELEVEN

Investigation of MicroRNA Alterations in Leukemias and Lymphomas

George Adrian Calin *and* Carlo Maria Croce

Contents

Department of Molecular Virology, Immunology and Medical Genetics and Comprehensive Cancer Center, The Ohio State University, Columbus, Ohio

Methods in Enzymology, Volume 427
ISSN 0076-6879, DOI: 10.1016/S0076-6879(07)27011-9

Abstract

Alterations in miRNA genes play a critical role in the pathophysiology of many, perhaps all, human cancers: cancer initiation and progression can involve altera-tions of microRNA genes (miRNAs) encoding small noncoding RNAs that can regulate gene expression. The main mechanism of microRNoma (defined as the full complement of microRNAs present in a genome) alteration in cancer cell seems to result in aberrant gene expression characterized by abnormal levels of expression for mature and/or precursor miRNA sequences in compari-son with the corresponding normal tissues. Loss or amplification of miRNA genes has been reported in a variety of cancers, and altered patterns of miRNA expres-sion may affect cell cycle and survival programs. The causes of the widespread differential expression of miRNA genes between malignant and normal cells can be explained by the genomic location of these genes in cancer-associated genomic regions, by epigenetic mechanisms as well as by alterations of members of the processing machinery. Germline and somatic mutations in miRNAs or polymorphisms in the mRNAs targeted by miRNAs may also con-tribute to cancer predisposition and progression. miRNAs expression profiling has been exploited to identify miRNAs potentially involved in the patho-genesis of human cancers and has allowed the identification of signatures associated with diagnosis, staging, progression, prognosis, and response to treatment of human tumors. Here we present a flowchart of principal steps to produce, analyze, and understand the biological significance of miRNA microarray data.

1. MicroRNA Alterations are Involved in the Initiation and Progression of Every Type of Human Cancer

During the last few years, we pioneered the idea that small noncoding RNAs, named microRNA genes (miRNAs), are involved in human tumor-igenesis in general and, particularly, that *miR-15a* and *miR-16-1* are altered in patients with chronic lymphocytic leukemia (CLL) (Calin *et al.*, 2002, 2004a). miRNAs are a group of small, noncoding RNA (ncRNA) molecules distinct from, but related to small, interfering RNAs (siRNAs) that have been identified in a variety of organisms (for a comprehensive review see Bartel [2004]). These small 19 to 24 nucleotide (nt) RNAs are transcribed as parts of longer molecules named pri–miRNA that can be several kilobases (kb) in length and processed in the nucleus into hairpin RNAs of 60 to 110 nt called precursor miRNA (pre–miRNA). In animals, single-stranded miRNA binds specific messenger RNA (mRNA) through sequences that are signifi-cantly, though not completely, complementary to the target mRNA, mainly to the 3' untranslated region (3' UTR) (Ambros, 2004). By a mechanism that is not fully characterized, the bound mRNA remains untranslated, result-ing in reduced levels of the corresponding protein; alternatively, the bound mRNA can be degraded, resulting in reduced levels of both the corresponding transcript and consequently the protein. The miRNAs have been involved in various biological processes, such as hematopoietic B–cell lineage fate (*miR-181*), B–cell survival (*miR-15a* and *miR-16-1*), cell proliferation control (*miR-125b* and *let-7*), brain patterning (*miR-430*), pancreatic cell insulin secretion (*miR-375*), and adipocyte development (*miR-143*) (for review, see Harfe [2005], Sevignani *et al.* [2006], and Di Leva *et al.* [2006]).

Initially identified in the most common form of adult leukemia, B–cell chronic lymphocytic leukemia CLL (Calin *et al.*, 2002), alterations of miRNAs have been detected in many types of human tumors by different groups (for reviews see Esquela-Kerscher and Slack [2006] and Calin and Croce [2006a]). The miRNAs were proposed to contribute to oncogenesis functioning as tumor suppressor genes (as is the case of *miR-15a* and *miR-16-1*) or as oncogenes (as is the case of *miR-155* or *miR17-92* cluster). The genomic abnormalities found to influence the activity of miRNAs are similar to those described to target protein-coding genes, including chromosomal rearrangements, genomic amplifications or deletions, and mutations. In a specific tumor, both abnormalities in protein coding genes and miRNAs can be identified (Calin and Croce, 2006b). Homozygous mutations or the combination of deletion plus mutation in miRNA genes are rare events (Calin *et al.*, 2005) and the functional consequences of heterozygous sequence variations of miRNAs in human cancers have not been demonstrated (Diederichs and Haber, 2006). Furthermore, the roles of polymorphisms in the complementary sites of target mRNAs in cancer patients (He *et al.*, 2005) or individuals with predisposition to other hereditary diseases (Abelson *et al.*, 2006) have just begun to be understood. Therefore, at the present time, the main mechanism of miRNoma (defined as the full complement of micro-RNAs present in a genome) alteration in cancer cells seems to be represented by aberrant gene expression, and characterized by abnormal levels of expression for mature and/or precursor miRNA sequences. The number of samples analyzed by miRNA microarrays passed the mark of 1000 at the end of year 2006 (Fig. 11.1) and three of the most important evolving themes of miRNA deregulation in human tumors are represented by (a) miRNA profiling is a new diagnostic tool for cancer patients, (b) miRNA profiling represents a prognostic tool and (c) miRNA alterations can represents a new cause of cancer predisposition (for comprehensive reviews on these themes, see Calin and Croce [2006a,b]).

2. GENOME-WIDE MICRORNA PROFILING BY MICROARRAY

Several developments for high-throughput miRNA profiling include the use of a bead-based flow cytometric technique (Lu *et al.*, 2005), the quantitative RT-PCR for precursor miRNA (Schmittgen *et al.*, 2004), and

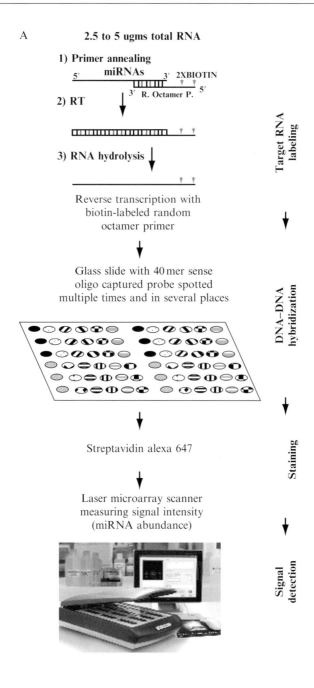

B

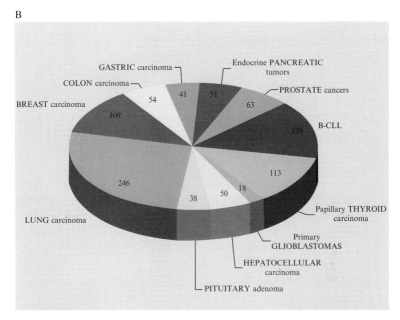

Figure 11.1 Principles of microarray technology used for microRNA profiling. (A) The microarray-based miRNA profiling is presented as described in the majority of profiling studies on primary tumors, initially developed by Liu *et al.* (2004). Target labeling, hybridization, staining, and signal detection are the four main technical steps (presented on the right side). The different replicates of the spots on the glass slide (presented as different types of gray grids) represent different oligonucleotide sequences corresponding to sequences from the precursor miRNA or active miRNA molecule. The main advantage of the microarray-based miRNA profiling is the high standardization of the procedure (allowing the processing of tens of samples in parallel), making it relatively easy to be performed. Several technical aspects regarding this technology can be found elsewhere (Liu *et al.*, 2007). (B) Profiling of 921 hematological and solid tissue primary cancers (and corresponding normal tissues samples) performed by the microarray technology developed by Liu *et al.* (2004) as published to date (numbers are according to data compiled in Calin and Croce [2006a] and from new reports of Bottoni *et al.* [2005] and Weber *et al.* [2006]). The general consensus is that profiling allowed the definitions of signatures associated with diagnosis and prognosis. Modified with permission from *Nature Reviews Cancer* (Calin, G. A., and Croce, C. M. [2006]. MicroRNA signatures in human cancers. *Nat. Rev. Cancer* **6**, 857–866.) Copyright (2006) MacMillan Magazines Ltd. (See color insert.)

active miRNA (Chen *et al.*, 2005; Raymond *et al.*, 2005) or the miRAGE, the genome-wide miRNA analysis with Serial Analysis of Gene Expression (SAGE) (Cummins *et al.*, 2006). The most commonly used high-throughput technique for the assessment of cancer-specific expression levels for hundreds of miRNAs in large numbers of samples is represented by oligonucleotide miRNA microarrays (see Fig. 11.1) (for reviews see Calin and Croce [2006a], Hammond [2006], and Kim and Nam [2006]). As miRNA

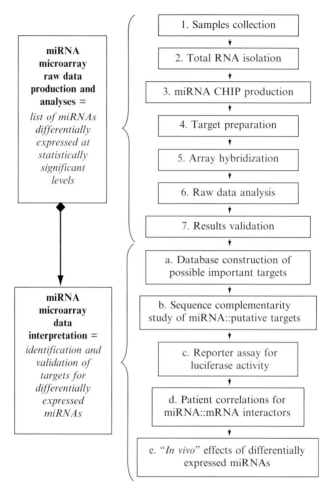

Figure 11.2   Flowchart of investigation of miRNA alterations in leukemias and lympho-
mas. The technical aspects and interpretation of results for the genome-wide miRNA
microarray investigation is presented in a logical flow. Here, we present on the left side the
two main tasks, and on the right side, the principal steps necessary to accomplish them.

profiling is one of the most active fields in miRNA research and the number
of published papers is exponentially growing, we will present the main parts
of the experimental design and technological steps that we used while
performing this type of study initially described in Liu *et al.* (2004)
(Fig. 11.2). Several technical aspects regarding this technology can be
found elsewhere (Liu *et al.*, 2007). Also, the definition of the main technical
terms used in miRNA profiling is presented in the glossary (Table 11.1).
Because at least 15 different miRNA microarray platforms are available, it is
a good idea before starting any experimental design to simply check online
for the one better suited for your needs.

Table 11.1 Glossary of terms related to the microRNA microarray research

MiRNoma: The full spectrum of miRNAs expressed in a particular cell type.

Interactor miRNA: A specific miRNA that is predicted by a computational program or has proved to directly interact with a specific messenger mRNA.

Target mRNA: The messenger RNA that contains a predicted site/sites of interaction with miRNAs or has proved to directly interact with a specific miRNA.

Gene–expression profiling: Determination of the level of expression of hundreds or thousand of genes through the use of microarrays. Total RNA extracted from the test tissue or cells and labeled with a fluorescent dye is tested for its ability to hybridize to the spotted nucleic acids.

Hierarchical clustering technique: A computational method that groups genes (or samples) into small clusters and then groups these clusters into increasingly higher level clusters. As a result, a dendrogram (i.e., tree) of connectivity emerges.

Prediction analysis of microarrays (PAM): A statistical technique that identifies a subgroup of genes that best characterizes a predefined class and uses this gene set to predict the class of new samples.

Significance analysis of microarrays (SAM): A statistical method used in microarray analyses that calculates a score for each gene and thus identifies genes with a statistically significant association with an outcome variable such as transfection with a specific miRNA.

GeneSpring software: GeneSpring is a powerful analysis tool that analyzes the scanned microarray data by assigning experiment parameters and interpretation to filter genes for differential expression and cluster to identify similar regulated groups.

MIAME: The Minimum Information About a Microarray Experiment that is needed to enable the interpretation of the results of the experiment unambiguously and potentially to reproduce the experiment. MIAMExpress (a MIAME-compliant microarray data submission tool) is currently available at http://www.ebi.ac.uk/miamexpress/.

Univariate vs. multivariate analyses: Univariate analysis explores each variable in a data set separately and describes the pattern of response to the variable. It describes each variable on its own. Multivariate statistical analysis describes a collection of procedures that involves observation and analysis of more than one statistical variable at a time. It describes all variables together.

Gene/Protein list: A group of genes/proteins with a common property, such as putative interaction with miRNAs or same expression profiles.

2.1. Samples collection

In our experience, the ratio of normal to tumor samples has to be approximately 1 to 10 (10% normal samples, e.g., 90 tumors and at least 10 normal correspondent controls, each from distinct individuals). The ideal situation is that of paired normal-tumor specimens from the same patient, as described in Yanaihara *et al.* (2006). It is a safe procedure to include few cell lines to be performed with the same batch of microarray slides. Although the cell lines are full of genetic abnormalities never identified in patient tumors, cell lines can be useful models to study the effects of down- or up-regulated miRNAs. It is useful from the start of experiment to select the clinical samples with different characteristics (e.g., colorectal cancer with and without microsatellite instability or CLL with bad and with good prognosis) in groups of relatively similar numbers (and as many as 30 for each group, to be able to further perform various statistical analyses, see following). If the normal controls are not enough to confer statistical power, the data can still be analyzed by comparing the two (or more) clinically distinct sets. As disease (in our specific case, cancer) oriented profiling is using few different types of samples (e.g., "malignant" and "normal controls" in the simplest instance) no replicates from each specific sample are needed. In the case of more detailed studies involving limited, but biologically very different types of samples (e.g., cells transfected with a specific reagent and nontransfected controls), three replicates of each sample are recommended.

2.2. Total RNA isolation

Total RNA isolation can be performed using the standard Tri-reagent or Trizol protocol (Molecular Research Center, Cincinnati, OH) or the newly developed kits for RNA extraction (also at higher prices per sample). If using RNA kit extraction, check carefully the protocol of the kit as some filters have a 200-bp RNA length cut-off-value, and therefore, both mature and precursor molecules will be retained in the filter. If the samples should be submitted to the microarray facility from long distances, a good option could be to resuspend the cell pellet in Trizol and ship in this way. As small RNAs are less stable in ethanol, resuspension in ethanol before shipment should be avoided. Regarding the amount of total RNA used, between 2.5 and 5 μg of total RNA are sufficient for each sample. For clinical materials available in small amount, other options could be the quantitative RT-PCR or microarray platforms specifically designed for very low input miRNA (Wang *et al.*, 2006). Regarding the quality of RNA, as usual, higher is better. In our experience, biologically significant microarray data can also be obtained from partially degraded total RNA, as small molecules are less degraded, but in such cases, several replicates of the same sample should be performed to assess reproducibility of results.

One important issue for the clinical samples is the use of paraffin-extracted RNA for the microarray experiments. In our experience this use is possible, and the most important variable for the quality is not, as supposed, the method of RNA extraction from the paraffin block, but the laboratory of provenience. We have performed a study (G. A. Calin, A. Scarpa, and C. M. Croce, unpublished data) in which, from the same GIST tumors (gastrointestinal stromal tumors), we have analyzed both frozen tissues and paraffin-stored tissues, and the results were quite similar, with better data obtained from frozen tissues extracted total RNA.

2.3. MicroRNA CHIP production and description

The technology described by Liu *et al.* (2004) used CodeLink arrays that are one-channel microarrays for expression studies, with common basic technical features to the miRNA chips described later. The current miRNA microarray used at the microarray facility at the Comprehensive Cancer Center at The Ohio State University (CCC-OSU) is version 3.0 (microRNACHIPv3), and contains probes against 578 precursor miRNA sequences (329 *Homo sapiens* and 249 *Mus musculus*), and 3 *Arabidopsis thaliana* as negative controls. These correspond to human and mouse miRNAs found in the microRNA registry at www.sanger.ac.uk/Software/Rfam/mirna/ (October, 2005) or collected from published papers (Calin *et al.*, 2004b). All the sequences were confirmed by BLAST alignment with the corresponding genome, and the hairpin structures were analyzed at www.bioinfo.rpi.edu/applications/mfold/old/rna. All 40-mer oligos were screened for their cross-homology to all genes of the relevant organism, number of bases in alignment to a repetitive element, amount of low-complexity sequence, maximum homopolymeric stretch, global and local G+C content, and potential hairpins (self 5-mers). The best oligo that contained each active site of each microRNA was selected. Next, we attempted to design for each specific miRNA gene an oligo that did not contain the active site, meaning an oligo that did not overlap the selected oligo(s) by more than 10 nt. To design each of these additional oligos, we required less than 75% global cross-homology and less than 20 bases in any 100% alignment to the relevant organism, less than 16 bases in alignments to repetitive elements, less than 16 bases of low-complexity, homopolymeric stretches of no more than six bases, G+C content between 30 and 70% and no more than 11 windows of size 10 with G+C content outside 30 to 70%, and no self 5-mers. In addition, we designed oligos for eight human tRNAs and seven mouse tRNAs, using similar design criteria. We selected a single oligo for each, with the exception of the human and mouse initiators Met-tRNA-i, for which we selected two oligos each, described by Liu *et al.* (2004). As a rule, the oligos named as "...No1" correspond to the active miRNA molecule, while the oligos named as "...No2" correspond to the precursor. Detailed information could be obtained from the supplemental Table 11.1 (Liu *et al.*, 2004) or directly from the microarray facility at CCC-OSU at www.osuccc.osu.edu/microarray.

Note that the platform that we use is able to discriminate the independent variations in expression for the mature versus precursor molecule for a specific miRNA. This is biologically important as consistent abnormal expression of the precursor miRNA, but not of the correspondent active molecule, is found in various types of cancers. This finding raises the possibility that the nonactive part of the miRNA molecule could have independent and yet unknown functions that could be important in tumorigenesis. Therefore, we consider that fractionation of total RNA (selection of the very small-sized RNAs, usually less than 60 bp) before performing the microarray is disadvantageous as, by losing the larger precursor molecules from the RNA pool, a significant portion of results is voluntary disregarded. For example, using a large-scale microarray analysis of 540 samples, including 363 from six of the most frequent human solid tumor types and 177 normal controls, we found that cancer cells showed distinct miRNA profiles compared with normal cells for both active and precursor molecules (Volinia *et al.*, 2006).

2.4. Target preparation

Two and one half to 5 μg of total RNA is separately added to a reaction mix in a final volume of 12 μl, containing 1 μg of [3′(N)8-(A)12-biotin-(A)12-biotin 5′] oligonucleotide primer. The total RNA and oligo primer mixture is incubated for 10 min at 70° for specific annealing first and chilled on ice. With the mixture remaining on ice, 4 μl of 5× first-strand buffer, 2 Vol. 0.1-M DTT, 1 μl of 10-mM dNTP mix, and 1-μl Superscript^{TM} II RNaseH⁻ reverse transcriptase (Invitrogen) (200 U/μl) is added to a final volume of 20 μl, and the mixture incubated for 90 min in a 37°water bath. After incubation for first-strand cDNA synthesis, 3.5 μl of 0.5-M NaOH/50-mM EDTA is added to 20 μl of first-strand reaction mix and incubated at 65° for 15 min to denature the RNA/DNA hybrids and degrade RNA templates. Then, 5 μl of 1-M Tris-HCI, pH 7.6 (Sigma), is added to neutralize the reaction mix and first-strand–labeled cDNA targets is stored in 28.5 μl at −80° until chip hybridization.

2.5. Array hybridization

Labeled first-strand cDNA targets from total RNA are used for hybridization on each OSU-CCC miRNA microarray. In particular cases, when the quality of RNA is very good and the material obtained from precious samples (such as long time clinically followed-up patient samples), we were able to obtain good quality and reproducible results by using as little as 1-μg total RNA. All probes on these microarrays are 5′ amine-modified 40-mer oligonucleotides spotted on a CodeLink activated slide (GE Healthcare,

Piscataway, NJ) by contacting technologies and covalently attached to a polymeric matrix. The microarrays are hybridized in 6× SSPE/30% form-amide at 25° for 18 h, washed in 0.75× TNT at 37° for 40 min, and processed using direct detection of the biotin-containing transcripts by Streptavidin-Alexa647 conjugate. Processed slides are scanned using an Axon 4000B Scanner (Molecular Device, Sunnyvale, CA) with the red laser set to 635 nm, at power 10% and PMT 800 setting, and a scan resolution of 10 μm.

2.6. Raw data analysis

Ideally, the same raw data should be analyzed by two distinct bioinformatics using two independent methods of analysis. In this way, we were able to find a unique miRNA signature associated with prognostic factors and-disease pro-gression in B-cell CLL (Calin et al., 2005). One strategy developed by Stefano Volinia is the following: Microarray images are analyzed by using GenePix Pro and postprocessing performed according to Volinia et al. (2006). Briefly, average values of the replicate spots of each miRNA are background-subtracted, normalized, and subjected to further analysis. miRNAs are retained when present in at least 20% of samples, (meaning the miRNAs are measured as present in at least the smallest group in the dataset) and when at least 20% of the miRNA had fold change of more than 1.5 from the gene median. Absent calls are threshold to 22 (4.5 in log2 scale) prior to statistical analysis. This level is the average minimum intensity level detected above background in miRNA chips experiments. Three different normalization procedures are assessed: global median, housekeeping, and cyclic loess. The SAM (Significance Analysis of Microarray, http://www-stat.stanford.edu/~tibs/SAM/index.html) and PAM (Prediction Analysis of Microarrays, http://www-stat.stanford.edu/~tibs/PAM/index.html) tests performed for the different classes (usually, "cancer" vs. "normal" or "cancer set 1" vs. "cancer set 2") over the respective expression tables are compared. Cyclic loess exhibited the best improvement in the reduction of variability and yielded the highest number of significant miRNAs, across the different tests, in agreement with an evaluation of different normalizations methods for CodeLink microarrays (Wu et al., 2005). Cyclic loess uses the MA plot and loess smoothing to estimate intensity-dependent differences in each pair of chips in the dataset, and then removes these differences by centering the loess line to zero. The procedure is iterated until intensity-dependent differences are removed from all arrays. Cyclic loess was implemented using the Bioconductor package/function affy/normalization loess. miRNA nomenclature was according to the miRNA database at Sanger Center. Differentially expressed miRNAs are identified by using the t test procedure included in SAM. SAM calculates a score for each gene on the basis of the change in expression relative to the standard deviation of all measurements. The miRNA signatures are determined by applying the nearest shrunken centroids. This method identifies a subgroup of genes that best characterizes two sample groups. The prediction error is calculated by means of tenfold cross-validation. Using this specific approach, we were able to identify a common miRNA expression signature of human solid tumors that defines cancer gene targets (Volinia *et al.*, 2006).

A second strategy developed by Manuela Ferracin and Massimo Negrini (Calin *et al.*, 2004a; Iorio *et al.*, 2005) consists of raw data normalization and analysis using GeneSpring® software (Silicon Genetics, Redwood City, CA). Expression data are median-centered using both the GeneSpring normalization option and Global Median normalization of the Bioconductor package (www.bioconductor.org). In our personal experience, no substantial difference was found. Statistical comparisons are done using the GeneSpring ANOVA tool, and the Support Vector Machine tool of Gene-Spring is used for the cross-validation and test-set prediction. In this way, the miRNAs able to best separate the groups are identified and confirmed by two different methods of raw data normalization. All data are submitted using MIAMExpress to the Array Express database at http://www.ebi.ac.uk/arrayexpress/. Using this specific strategy, we were able to find a common signature for breast cancer (Iorio *et al.*, 2005) as well as, for the first time, a miRNA signature of hypoxia (Kulshreshtha *et al.*, 2007).

The end results of raw data analyses consist of (1) dendrograms showing the clustering of multiple samples according to the transfection status and type of transfectants (e.g., miRNAs, scrambled oligos, or empty vectors) and (2) gene lists containing genes differentially expressed at high statistical significance ($P < 0.05$ or, better, $P < 0.01$). The level of differential expression that can be considered biologically significant represents one key issue. As fold differences of 10 to 20 are found to be important, we support a view in which much smaller changes ($>$ twofold) could also be significant. The explanation resides in the multiplicity of targets of a specific miRNA and the large number of altered miRNAs, making it likely that two or more target protein-coding genes on the same pathway are consequently disturbed. An opposite view is that in some cases the described changes in miRNA expression between normal and tumor cells could be biologically irrelevant: the main interactions with various targets could have not additive, but antagonistic, biological consequences (e.g., repression of both proapoptotic and antiapoptotic genes) (Calin and Croce, 2006a).

2.7. Validation of microRNA results

Confirmation of microarray data is done by Northern blots and quantitative real-time RT-PCR analysis, such as the assay for active miRNA (Chen *et al.*, 2005) or the quantitative RT-PCR for pre-miRNA (Schmittgen *et al.*, 2004). We, and others, were able to prove that our approaches of data analysis are extensively confirmed (Calin *et al.*, 2004a, 2005; Iorio *et al.*, 2005; Liu *et al.*, 2004; Roldo *et al.*, 2006; Yanaihara *et al.*, 2006). For Northern blot analysis, 10 to 20 μg of total RNA is used for each sample to run on a 15% polyacrylamide denaturing (urea) criterion precast gel (Bio-Rad) and then transferred onto Hybond-N+ membrane (Amersham Pharmacia Biotech). The blots are performed as described (Calin *et al.*, 2002). Quantitative RT-PCR for miRNA precursors and active molecules are performed as described for pre-miRNAs or for active molecule of miRNA. As the miRNA signatures are extensive, perform Northern blot (15% TBE-Urea gels) for two or three differentially expressed miRNAs, and other interesting miRNAs can be confirmed by quantitative RT-PCR. The confirmation rate for the miRNA CHIP is higher as 80% for the majority of experiments analyzed up-to-date. The batch of total RNA used for confirmation has to be the same as that used for microarray data.

As the final output of the previous steps is simply represented by a list of miRNA genes differentially expressed at certain folds and with a certain probability, it is difficult for the biologists or clinicians involved in the study should immediately determine a meaning of these data. Therefore, to understand the data significance, we consider that at least one distinct step should be performed (see Fig. 11.2). This is represented by the identification of targets of differentially expressed miRNA and by the analyses of the biological consequences of reexpressing down-regulated miRNA or shutting down overexpressed miRNAs.

3. IDENTIFICATION AND VALIDATION OF TARGETS FOR DIFFERENTIALLY EXPRESSED miRNAs

Some miRNAs can down-regulate large numbers of target mRNAs (Lim *et al.*, 2005), and it is estimated that miRNAs could regulate ~30% of the human genome (Bartel, 2004). miRNAs seem to be responsible for fine regulation of gene expression, "tuning" the cellular phenotype during delicate processes like development and differentiation in all organisms, from plants to mammals (Sevignani *et al.*, 2006). Target identification has been hampered by the fact that in animals, in contrast to plants, miRNAs do not bind perfectly to their targets. A few nucleotides typically remain unbound, yielding complex secondary structures. To date, only a few miRNA::mRNA interactions with importance for cancer pathogenesis have been proven (e.g., *let-7* and RAS oncogenes, *miR-15* and *miR-16* and BCL2 oncogene *miR-17-5p* and *miR-20a* and the tumor suppressor E2F1 transcription factor or *miR-127* and the oncogene BCL6) (Cimmino *et al.*, 2005; Johnson *et al.*, 2005; O'Donnell *et al.*, 2005; Saito *et al.*, 2006). Several different target prediction programs or methods were developed in a span of less than 5 years (for a comprehensive review, see Rajewsky [2006]).

Each has its strengths and weaknesses, but as a rule, the set of putative targets for a specific miRNA is quite different according to each program used, making it difficult to select false positives. We mainly used the following four computational prediction programs for miRNA targets: DianaMicroT (Kiriakidou *et al.*, 2004), TargetScan (Lewis *et al.*, 2005), miRanda (Enright *et al.*, 2003), and PicTar (Krek *et al.*, 2005). A clear advance was the addition of the miRGen software (at http://www.diana.pcbi.upenn.edu/miRGen.html): its targets interface provides access to unions and intersections of the four widely used target prediction programs, and experimentally supported targets from TarBase (Megraw *et al.*, 2006).

We will present our approach for target identification used in the analyses of the identification of important biological targets for the CLL miRNAs signature. Using this method, we were able to find interactions also predicted by several programs (such as *miR-15* and *miR-16* and BCL2) (Cimmino *et al.*, 2005), predicted by few programs (such as *miR-29* and *miR-181* and TCL1) (Pekarsky *et al.*, 2006), or not predicted yet by any program (the interactor miRNAs with ZAP-70 protein) (G. A. Calin and C. M. Croce, manuscript in preparation). All these are of crucial significance for the understanding of CLL pathogenesis, as BCL2 is the main apoptotic player in the malignant B cells from these patients, TCL1 is overexpressed specifically in the patients with a negative prognosis that have a significantly shorter survival, and ZAP-70 expression is one of the main prognostic factors in CLL (Chiorazzi *et al.*, 2005; Kipps, 2001). The principal steps are presented in Fig. 11.2.

3.1. Database construction of biologically important possible targets

The starting point was the carefully construction of a database of genes/proteins important for the B-cell tumorigenesis and for initiation and development of CLL. It contains two different types of transcripts: (1) protein-coding genes reported to be important for the biology of CLL and (2) a large number of noncoding RNAs (ncRNAs), including the precursor miRNAs for all known miRNAs. We previously showed that this type of *in silico* analysis is able to link the human miRNAs to fragile sites and genomic regions involved in cancer (Calin *et al.*, 2004b) and that independent laboratories have experimentally proven our predictions (Johnson *et al.*, 2005; Zhang *et al.*, 2006). Because new genes identification, as proved by the discovery of several families of small RNA genes, such as piRNAs (Kim, 2006) and the 21U-RNAs (Ruby *et al.*, 2006), is an active field, a systematic adjustment of the database and screen for new putative interactions is mandatory. Regarding the protein-coding genes reported to be important for the biology of CLL, these were selected after a PubMed search of published reports. Also, the personal experience achieved on the field of

CLL in the past and sharing prepublication data from world-renowned experts in the study of leukemia could be of strong support. Regarding the noncoding RNAs, the selection was done in various ways. First, we used available databases, such as the regulatory noncoding RNAs database at http://biobases.ibch.poznan.pl/ncRNA/ (Szymanski et al., 2006) or NON-CODE at http://noncode.bioinfo.org.cn/. Such ncRNA databases are intended to provide information on the sequences and functions of transcripts that do not code for proteins, but perform regulatory roles in the cell. The sequences included in the databases have been at least partially characterized in terms of their function or expression and therefore have the highest probability to be targets of different mechanisms of gene regulation. Second, we used the miRBase::Sequences in miRNA registry at http://microrna.sanger.ac.uk/sequences/index, which is the main miRNA online resource (for description, see Griffiths-Jones et al. [2006]) to find the pre-miRNA sequences of all cloned human miRNAs. In this way, we are using the maximum amount of information and number of transcripts to be screened by computational methods. As information regarding the biological significance of such ncRNA in human diseases is practically absent, we did not use PubMed literature screening as an inclusion/exclusion criteria, but only as a ranking criteria for importance in CLL. Per se, finding an interaction miRNA::ncRNA is a major advance in better understanding miRNA function, and therefore an important goal is to demonstrate that such interactions exist and have pathogenetic and maybe clinical significance. At the final point, we produced a list of 50 to 100 genes/proteins known or supposed to be involved in the tumor type(s) or pathway(s) or process(es) associated with CLL pathogenesis.

3.2. Study of sequence complementarities

Generally, there are two different ways to use the target prediction software: (1) starting from a specific miRNA and giving a rank to each prediction; and (2) starting from a specific candidate target and identifying all the possible miRNA interactors. In this way, we identified several thousands of predicted targets (e.g., only TargetScanS is predicting for the family *miR-15a*/16-1/195 more than 400 coding mRNA). By starting from a large, but specific, list of candidate targets with biological importance the number of false positives is significantly reduced. By intersecting our database with the various computational programs, we found a relatively small number of positive hits (about 20, including BCL2 and TCL1, but not ZAP-70). An additional step was to perform an in-house complementarity search using various methods, such as automated FASTA by using Perl scripts or alignment programs such as Sequencer (GeneCode Corporation). Perl scripts for the automatic submission of blast jobs and for the retrieval of the search results are based on the LPW, HTML, and HTPP Perl modules and BioPerl modules (Calin *et al.*, 2004b). In this way, all the possible alignments were screened, and such an approach gives us the pairing between an miRNA located in a very frequently deleted chromosomal region in CLL and ZAP-70 that we also experimentally confirmed. Because target prediction programs are constantly improving and other algorithms are developed, we are performing a systematic rescan of our database.

3.3. Reporter assays for luciferase activity

The following is a well-established, "easy-to-do" technique that in our hands (Cimmino *et al.*, 2005) and others' (Lewis *et al.*, 2003) was very successful in defining the direct interaction miRNA::mRNA. It simply consists of measuring the luciferase activity after reporter plasmids containing the imperfect complementarity region of interactor mRNA and the miRNAs of interest (putative interactor miRNA) are cotransfected in recipient cells, such as MEG-01, 293 HEK cells, or HeLA cells. In our experience, MEG-01 are highly transfectable cells (efficiency about 99% for various types of vectors); furthermore, MEG-01, a chronic leukemia-derived cell line has no expression of *miR-15a* and *miR-16-1* genes because of the deletion of one allele and alteration of the other *miR-15a/16-1* locus on chromosome 13q14.3. Therefore, being a natural knockout cell for these miRNAs, MEG-01 represented a suited system to analyze the functional importance of target interactions as we have proved for BCL2::miR-15/16 interaction (Cimmino *et al.*, 2005). Briefly, a $3'$UTR segment or a segment of the ncRNA transcript is amplified by PCR from human genomic DNA and inserted into the pGL3 control vector (Promega), using the XbaI site immediately downstream from the stop codon of luciferase. We usually also generate inserts with deletions from the site of perfect complementarity. Wild-type and mutant inserts are confirmed by sequencing. The cells are cotransfected in 12-well plates using siPORT neoFX (Ambion) according to the manufacturer's protocol and using 0.4 μg of the firefly luciferase report vector and 0.08 μg of the control vector containing Renilla luciferase, pRL-TK (Promega). For each well, 10 nM of *miR*-sense or anti-*miR*-precursor miRNA inhibitor (generally from Ambion, Pharmacon, or Fidelity) are used. The anti-miRNAs are antisense products of the same gene and are used as negative controls for the experiments. Firefly and Renilla luciferase activities are measured consecutively using the Dual-luciferase assays (Promega) 24 h after transfection.

3.4. Patient correlations for miRNA::target mRNA interactors expression

An important interaction between an miRNA and a biologically important target could be "direct" (sequence complementarity), but also "indirect" (e.g., via a direct transcription factor influenced by the miRNA). Therefore, in our opinion, this step is the most important, as it could also identify the indirect interactions involved in the mechanism of a specific disease. The best way to start is to confirm the correlation using proteins from the same lot of patients used for expression profiling. If found, check the same correlation in a transfected cell line with the miRNA of interest (sense plus antisense as control), which express the protein considered putative target. The sense and should have to have opposite effects. Furthermore, as both transcriptional and translational effects could be found, it is easier to start the correlation by simple semiquantitative RT-PCR. This was the way in which we published in the first paper linking miRNAs and cancer— the first-ever correlation study for miRNA and putative target mRNA expression that found the RARS gene (arginyl-tRNA synthetase) as a transcriptional target of *miR-16* (Calin *et al.*, 2002). We were able to identify by EST microarray in *miR-16* transfected leukemic MEG01 significant down-regulation of RARS with respect to a cell transfected with empty vector at very high significant values ($P < 0.0001$) (Cimmino *et al.*, manuscript in preparation). Also, keep in mind that, because only negative correlations have been proven up-to-date (miRNA inhibits transcription or translation), this doesn't mean the positive correlations did not exist. One proof in this sense is the identification of *miR-122*-positive influence on replication of hepatitis C virus (HCV) by interacting with the $5'$ noncoding region of the virus (Jopling *et al.*, 2005).

3.5. Identification of *in vivo* effects of differentially expressed miRNAs

Among the final steps in the interpretation of any miRNA microarray experiments should be the direct identification of *in vivo* effects as a consequence on miRNA::mRNA interaction. We were able to prove not only the direct interaction between *miR-15* and *miR-16* and BCL2, but also to show that BCL2 repression by these two miRNAs induces apoptosis in MEG01 leukemia cells by the activation of intrinsic apoptosis APAF1–caspase9–PARP pathway (Cimmino *et al.*, 2005). The transfected cells with the miRNA of interest have to be analyzed for at least two phenotypic effects: (1) the effects on cell proliferation and death, morphology and differentiation using flow-cytometry and/or Western blotting; and (2) the altered expression of the identified targets and the protein expression from the same or related pathways. The cells could be transfected with miRNA oligos or with plasmids constructs containing the miRNA of interest. As recently was proven that not only RNA polymerase II but also Pol III transcribes human miRNAs (Borchert *et al.*, 2006), both types of constructs (with Pol II or with Pol III promoters) can be used. Two observations from our experience and from putting together data from the published papers of others should be pointed out. First, in the majority of the miRNA studies reporting biological effects, the range of effects variation is less than

50% (e.g., see Chen *et al.* [2004] or Poy *et al.* [2004]). Such variations, for a normal protein-coding gene, are easily considered nonsignificant. One explanation can be the redundancy of miRNAs effects from the same family (same active sequence): we hypothesize that we have to transfect all the members of that family of miRNAs (genes with same active molecule sequence or few nucleotides differences) to obtain the full range and the full levels of effects. Therefore, the experiments have to be performed several times with adequate controls each time. Second, always extract RNA *and* proteins from the transfection experiments—do not discard if no "visible" effects can be identified. MiRNA effects are both at transcriptional as well as translational (post-transcriptional) levels. Use the RNA for EST microarrays (up to 30% of our genome transcripts seems to be regulated by miRNAs [Lim *et al.*, 2005]) and proteins for two-dimensional (2D) gels or WB. It will be useful for target identification. Also, keep in mind that an "oncogene" miRNA (amplified/overexpressed in tumors) can also be a "tumor suppressor" miRNA, depending on the critical targets in the relevant tissue. For example, the same miRNAs can target an oncogene in a particular tissue and a tumor suppressor protein-coding gene in a different tissue.

ACKNOWLEDGMENTS

Dr. Croce is supported by Program Project Grants from the National Cancer Institute and Dr. Calin by a Kimmel Foundation Scholar award and by the CLL Global Research Foundation. We acknowledge Dr. Chang-gong Liu for the miRNA microarray production and Drs. Massimo Negrini, Stefano Volinia, Manuela Ferracin, and Cristian Taccioli for the statistical analyses of miRNAs microarray data. We thank Dr. Muller Fabbri for the critical reading of the manuscript. We apologize to many colleagues whose work was not cited due to space limitations.

REFERENCES

Abelson, J. F., Kwan, K. Y., O'Roak, B. J., Baek, D. Y., Stillman, A. A., Morgan, T. M., Mathews, C. A., Pauls, D. L., Rasin, M. R., Gunel, M., Davis, N. R., Ercan-Sencicek, A. G., *et al.* (2006). Sequence variants in SLITRK1 are associated with Tourette's syndrome. *Science* **310,** 317–320.
Ambros, V. (2004). The functions of animal microRNAs. *Nature* **431,** 350–355.
Bartel, D. P. (2004). MicroRNAs: Genomics, biogenesis, mechanism, and function. *Cell* **116,** 281–297.
Borchert, G. M., Lanier, W., and Davidson, B. L. (2006). RNA polymerase III transcribes human microRNAs. *Nat. Struct. Mol. Biol.* **13,** 1097–1101.
Bottoni, A., Piccin, D., Tagliati, F., Luchin, A., Zatelli, M. C., and degli Uberti, E. C. (2005). miR-15a and miR-16-1 down-regulation in pituitary adenomas. *J. Cell Physiol.* **204,** 280–285.
Calin, G. A., and Croce, C. M. (2006a). MicroRNA signatures in human cancers. *Nat. Rev. Cancer* **6,** 857–866.

Calin, G. A., and Croce, C. M. (2006b). MicroRNA-cancer connection: The beginning of a new tale. *Cancer Res.* **66,** 7390–7394.

Calin, G. A., Dumitru, C. D., Shimizu, M., Bichi, R., Zupo, S., Noch, E., Aldler, H., Rattan, S., Keating, M., Rai, K., Rassenti, L., Kipps, T., *et al.* (2002). Frequent deletions and down-regulation of micro- RNA genes miR15 and miR16 at 13q14 in chronic lymphocytic leukemia. *Proc. Natl. Acad. Sci. USA* **99,** 15524–15529.

Calin, G. A., Ferracin, M., Cimmino, A., Di Leva, G., Shimizu, M., Wojcik, S. E., Iorio, M. V., Visone, R., Sever, N. I., Fabbri, M., Iuliano, R., *et al.* (2005). A MicroRNA signature associated with prognosis and progression in chronic lymphocytic leukemia. *N. Engl. J. Med.* **353,** 1793–1801.

Calin, G. A., Liu, C. G., Sevignani, C., Ferracin, M., Felli, N., Dumitru, C. D., Shimizu, M., Cimmino, A., Zupo, S., Dono, M., Dell'Aquila, M. L., Alder, H., *et al.* (2004a). MicroRNA profiling reveals distinct signatures in B cell chronic lymphocytic leukemias. *Proc. Natl. Acad. Sci. USA* **101,** 11755–11760.

Calin, G. A., Sevignani, C., Dumitru, C. D., Hyslop, T., Noch, E., Yendamuri, S., Shimizu, M., Rattan, S., Bullrich, F., Negrini, M., and Croce, C. M. (2004b). Human microRNA genes are frequently located at fragile sites and genomic regions involved in cancers. *Proc. Natl. Acad. Sci. USA* **101,** 2999–3004.

Chen, C., Ridzon, D. A., Broomer, A. J., Zhou, Z., Lee, D. H., Nguyen, J. T., Barbisin, M., Xu, N. L., Mahuvakar, V. R., Andersen, M. R., Lao, K. Q., Livak, K. J., *et al.* (2005). Real-time quantification of microRNAs by stem-loop RT-PCR. *Nucleic Acids Res.* **33,** e179.

Chen, C. Z., Li, L., Lodish, H. F., and Bartel, D. P. (2004). MicroRNAs modulate hematopoietic lineage differentiation. *Science* **303,** 83–86.

Chiorazzi, N., Rai, K. R., and Ferrarini, M. (2005). Chronic lymphocytic leukemia. *N. Engl. J. Med.* **352,** 804–815.

Cimmino, A., Calin, G. A., Fabbri, M., Iorio, M. V., Ferracin, M., Shimizu, M., Wojcik, S. E., Aqeilan, R., Zupo, S., Dono, M., Rassenti, L., Alder, H., *et al.* (2005). miR-15 and miR-16 induce apoptosis by targeting BCL2. *Proc. Natl. Acad. Sci. USA* **102,** 13944–13949.

Cummins, J. M., He, Y., Leary, R. J., Pagliarini, R., Diaz, L. A. J., Sjoblom, T., Barad, O., Bentwich, Z., Szafranska, A. E., Labourier, E., Raymond, C. K., Roberts, B. S., *et al.* (2006). The colorectal microRNAome. *Proc. Natl. Acad. Sci. USA* **103,** 3687–3692.

Diederichs, S., and Haber, D. A. (2006). Sequence variations of microRNAs in human cancer: Alterations in predicted secondary structure do not affect processing. *Cancer Res.* **66,** 6097–6104.

Di Leva, G., Calin, G. A., and Croce, C. M. (2006). MicroRNAs: Fundamental facts and involvement in human diseases. *Birth Defects Res. C Embryo Today* **78,** 180–189.

Enright, A. J., John, B., Gaul, U., Tuschl, T., Sander, C., and Marks, D. S. (2003). MicroRNA targets in *Drosophila*. *Genome Biol.* **5,** R1.

Esquela-Kerscher, A., and Slack, F. J. (2006). Oncomirs—microRNAs with a role in cancer. *Nat. Rev. Cancer* **6,** 259–269.

Griffiths-Jones, S., Grocock, R. J., van Dongen, S., Bateman, A., and Enright, A. J. (2006). miRBase: microRNA sequences, targets and gene nomenclature. *NAR Database Issue*, D140–D144.

Hammond, S. M. (2006). MicroRNA detection comes of age. *Nat. Methods* **3,** 12–13.

Harfe, B. D. (2005). MicroRNAs in vertebrate development. *Curr. Opin. Genet. Dev.* **15,** 410–415.

He, H., Jazdzewski, K., Li, W., Liyanarachchi, S., Nagy, R., Volinia, S., Calin, G. A., Liu, C. G., Franssila, K., Suster, S., Kloos, R. T., Croce, C. M., *et al.* (2005). The role of microRNA genes in papillary thyroid carcinoma. *Proc. Natl. Acad. Sci. USA* **102,** 19075–19080.

Iorio, M. V., Ferracin, M., Liu, C. G., Veronese, A., Spizzo, R., Sabbioni, S., Magri, E., Pedriali, M., Fabbri, M., Campiglio, M., Ménard, S., Palazzo, J. P., et al. (2005). MicroRNA gene expression deregulation in human breast cancer. Cancer Res. **65,** 7065–7070.

Johnson, S. M., Grosshans, H., Shingara, J., Byrom, M., Jarvis, R., Cheng, A., Labourier, E., Reinert, K. L., Brown, D., and Slack, F. J. (2005). RAS is regulated by the let-7 microRNA family. Cell **120,** 635–647.

Jopling, C. L., Yi, M., Lancaster, A. M., Lemon, S. M., and Sarnow, P. (2005). Modulation of hepatitis C virus RNA abundance by a liver-specific MicroRNA. Science **309,** 1577–1581.

Kim, V. N. (2006). Small RNAs just got bigger: Piwi-interacting RNAs (piRNAs) in mammalian testes. Genes Dev. **20,** 1993–1997.

Kim, V. N., and Nam, J. W. (2006). Genomics of microRNA. Trends Genet. **22,** 165–173.

Kipps, T. J. (2001). Chronic lymphocytic leukemia and related diseases. In "Williams hematology" (E. Beutler, M. A. Lichtman, B. S. Coller, T. J. Kipps, and U. Seligson, eds.), pp. 1163–1194. McGraw-Hill, New York.

Kiriakidou, M., Nelson, P. T., Kouranov, A., Fitziev, P., Bouyioukos, C., Mourelatos, Z., and Hatzigeorgiou, A. (2004). A combined computational-experimental approach predicts human microRNA targets. Genes Dev. **18,** 1165–1178.

Krek, A., Grun, D., Poy, M. N., Wolf, R., Rosenberg, L., Epstein, E. J., MacMenamin, P., da Piedade, I., Gunsalus, K. C., Stoffel, M., and Rajewsky, N. (2005). Combinatorial microRNA target predictions. Nat Genet. **37,** 495–500.

Kulshreshtha, R., Ferracin, M., Wojcik, S. E., Garzon, R., Alder, H., Agosto-Perez, F. J., Davuluri, R., Liu, C.-G., Croce, C. M., Negrini, M., Calin, G. A., and Ivan, M. (2007). A microRNA signature of hypoxia. Mol. Cell. Biol. in press.

Lewis, B. P., Burge, C. B., and Bartel, D. P. (2005). Conserved seed pairing, often flanked by adenosines, indicates that thousands of human genes are microRNA targets. Cell **120,** 15–20.

Lewis, B. P., Shih, I. H., Jones-Rhoades, M. W., Bartel, D. P., and Burge, C. B. (2003). Prediction of mammalian microRNA targets. Cell **115,** 787–798.

Lim, L. P., Lau, N. C., Garrett-Engele, P., Grimson, A., Schelter, J. M., Castle, J., Bartel, D. P., Linsley, P. S., and Johnson, J. M. (2005). Microarray analysis shows that some microRNAs downregulate large numbers of target mRNAs. Nature **433,** 769–773.

Liu, C. G., Calin, G. A., Meloon, B., Gamliel, N., Sevignani, C., Ferracin, M., Dumitru, C. D., Shimizu, M., Zupo, S., Dono, M., Alder, H., Bullrich, F., et al. (2004). An oligonucleotide microchip for genome-wide microRNA profiling in human and mouse tissues. Proc. Natl. Acad. Sci. USA **101,** 9740–9744.

Liu, C.-G., Liu, X., and Calin, G. A. (2007). MicroRNoma genome-wide profiling by microarray. In "Bioinformatics: A Practical Approach" (S.-Q. Ye, ed.). Vol. in press. Taylor and Francis Group, LLC.

Lu, J., Getz, G., Miska, E. A., Alvarez-Saavedra, E., Lamb, J., Peck, D., Sweet-Cordero, A., Ebert, B. L., Mak, R. H., Ferrando, A. A., Downing, J. R., Jacks, T., et al. (2005). MicroRNA expression profiles classify human cancers. Nature **435,** 834–838.

Megraw, M., Sethupathy, P., Corda, B., and Hatzigeorgiou, A. G. (2006). miRGen: A database for the study of animal microRNA genomic organization and function. Nucleic Acids Res. 2006 Nov 15; [Epub ahead of print].

O'Donnell, K. A., Wentzel, E. A., Zeller, K. I., Dang, C. V., and Mendell, J. T. (2005). c-Myc-regulated microRNAs modulate E2F1 expression. Nature **435,** 839–843.

Pekarsky, Y., Santanam, U., Cimmino, A., Palamarchuk, A., Efanov, A., Maximov, V., Volinia, S., Alder, H., Liu, C. G., Rassenti, L., Calin, G., Hagan, J. P., et al. (2006). Tcl1 expression in CLL is regulated by miR-29 and miR-181. Cancer Res. In press.

Poy, M. N., Eliasson, L., Krutzfeldt, J., Kuwajima, S., Ma, X., Macdonald, P. E., Pfeffer, S., Tuschl, T., Rajewsky, N., Rorsman, P., and Stoffel, M. (2004). A pancreatic islet-specific microRNA regulates insulin secretion. *Nature* **432,** 226–230.

Rajewsky, N. (2006). MicroRNA target predictions in animals. *Nat. Genet.* S8–S13.

Raymond, C. K., Roberts, B. S., Garrett-Engele, P., Lim, L. P., and Johnson, J. M. (2005). Simple, quantitative primer-extension PCR assay for direct monitoring of microRNAs and short-interfering RNAs. *RNA* **11,** 1737–1744.

Roldo, C., Missiaglia, E., Hagan, J. P., Falconi, M., Capelli, P., Bersani, S., Calin, G. A., Volinia, S., Liu, C. G., Scarpa, A., and Croce, C. M. (2006). MicroRNA expression abnormalities in pancreatic endocrine and acinar tumors are associated with distinctive pathological features and clinical behavior. *J. Clin. Oncol.* **24,** 4677–4684.

Ruby, J. G., Jan, C., Player, C., Axtell, M. J., Lee, W., Nusbaum, C., Ge, H., and Bartel, D. P. (2006). Large-scale sequencing reveals 21U-RNAs and additional micro-RNAs and endogenous siRNAs in *C. elegans. Cell* **127,** 1193–1207.

Saito, Y., Liang, G., Egger, G., Friedman, J. M., Chuang, J. C., Coetzee, G. A., and Jones, P. A. (2006). Specific activation of microRNAs-127 with downregulation of the proto-oncogene BCL6 by chromatin-modifying drugs in human cancer cells. *Cancer Cell* **9,** 435–443.

Schmittgen, T. D., Jiang, J., Liu, Q., and yang, L. (2004). A high-throughput method to monitor the expression of microRNA precursor. *Nucleic Acid Res.* **32,** 43–53.

Sevignani, C., Calin, G. A., Siracusa, L. D., and Croce, C. M. (2006). Mammalian micro-RNAs: A small world for fine-tuning gene expression. *Mamm. Genome* **17,** 189–202.

Szymanski, M., Erdmann, V. A., and Barciszewski, J. (2006). Noncoding RNAs database (ncRNAdb). *Nucleic Acids Res.* Dec 14; [Epub ahead of print].

Volinia, S., Calin, G. A., Liu, C. G., Ambs, S., Cimmino, A., Petrocca, F., Visone, R., Iorio, M., Roldo, C., Ferracin, M., Prueitt, R. L., Yanaihara, N., *et al.* (2006). A microRNA expression signature of human solid tumors defines cancer gene targets. *Proc. Natl. Acad. Sci. USA* **103,** 2257–2261.

Wang, H., Ach, R. A., and Curry, B. (2006). Direct and sensitive miRNA profiling from low-input total RNA. *RNA* **13,** 151–159.

Weber, F., Teresi, R. E., Broelsch, C. E., Frilling, A., and Eng, C. (2006). A limited set of human MicroRNA is deregulated in follicular thyroid carcinoma. *J. Clin. Endocrinol. Metab.* **91,** 3584–3591.

Wu, W., Dave, N., Tseng, G. C., Richards, T., Xing, E. P., and Kaminski, N. (2005). Comparison of normalization methods for CodeLink Bioarray data. *BMC Bioinformatics.* **6,** 309.

Yanaihara, N., Caplen, N., Bowman, E., Kumamoto, K., Okamoto, A., Yokota, J., Tanaka, T., Calin, G. A., Liu, C. G., Croce, C. M., and Harris, C. C. (2006). microRNA signature in lung cancer diagnosis and prognosis. *Cancer Cell* **9,** 189–198.

Zhang, L., Huang, J., Yang, N., Greshock, J., Megraw, M. S., Giannakakis, A., Liang, S., Naylor, T. L., Barchetti, A., Ward, M. R., Yao, G., Medina, A., *et al.* (2006). Micro-RNAs exhibit high frequency genomic alterations in human cancer. *Proc. Natl. Acad. Sci. USA* **103,** 9136–9141.

CHAPTER TWELVE

DISCOVERY OF PATHOGEN-REGULATED SMALL RNAS IN PLANTS

Surekha Katiyar-Agarwal *and* Hailing Jin

Contents

Abstract

Small RNAs have emerged as one of the most important regulators for gene expression in eukaryotes. Small RNA-mediated gene silencing has been shown to play an essential role in antiviral defense in both plant and animal systems (Li and Ding, 2005; Voinnet, 2005; Wang *et al.*, 2006). These viral RNA-generated small interfering RNAs (siRNAs) are extragenomic in origin. Studies from our lab and others suggest that host-endogenous small RNAs also play an important role in plant defense in response to other pathogens besides viruses (Katiyar-Agarwal *et al.*, 2006; Navarro *et al.*, 2006). The methods described here provide an opportunity to identify many more novel pathogen-regulated small RNAs in plants, which will help in understanding the regulatory mechanism of plant immunity. Here, we introduce the approaches of powerful high-throughput parallel sequencing and hybridization-based technologies for the discovery and detection of pathogen-regulated small RNAs. We mainly compare and discuss the methods of low–molecular-weight (LMW) RNA extraction from pathogen-infected tissue and strategies for detecting endogenous small RNAs by Northern blot analysis.

Department of Plant Pathology, Center for Plant Cell Biology and Institute for Integrative Genome Biology, University of California—Riverside, Riverside, California

Methods in Enzymology, Volume 427
ISSN 0076-6879, DOI: 10.1016/S0076-6879(07)27012-0

1. Introduction

Small RNA-mediated gene silencing has emerged as one of the most important regulatory mechanisms for gene expression in eukaryotes. Small RNAs are a population of 20- to 30-nt noncoding RNAs that regulate gene expression by guiding mRNA cleavage, translational inhibition, or chromatin modification (Baulcombe, 2005; Sontheimer and Carthew, 2005). A highly complex and abundant small RNA population has been discovered in *Arabidopsis* by use of deep-parallel sequencing technologies, including massively parallel signature sequencing (MPSS) and 454 sequencing (Brenner *et al.*, 2000; Henderson *et al.*, 2006; Lu *et al.*, 2005, 2006; Rajagopalan *et al.*, 2006; Margulies *et al.*, 2005).

Small RNAs can be divided into two classes: microRNAs (miRNAs) and siRNAs. miRNAs are generated from long hairpin precursors and are evolutionarily conserved across species (Ambros, 2003). In *Arabidopsis,* more than 100 miRNAs have been reported (Jones-Rhoades *et al.*, 2006; Mallory and Vaucheret, 2006). Many are highly abundant and can be easily detected (Mallory and Vaucheret, 2006). Many miRNAs are involved in plant development and hormone signaling. Some also play important roles in response to environmental stresses and nutrient deprivation (Fujii *et al.*, 2005; Jones-Rhoades and Bartel, 2004; Sunkar and Zhu, 2004; Sunkar *et al.*, 2006). One miRNA was shown to contribute to basal defense against bacteria by regulating auxin signaling (Navarro *et al.*, 2006). The endogenous siRNAs are generated from double-stranded RNA as a result of antisense transcription or the activity of cellular RNA-dependent RNA polymerases (RDRs) (Plasterk, 2002; Waterhouse *et al.*, 2001).

siRNAs are extremely diverse and normally not conserved across species. Most are less abundant than miRNAs and therefore are difficult to detect. Although endogenous siRNAs greatly outnumber miRNAs in plants, their functional roles are still largely unknown, except for trans-acting siRNAs in plant development and hormone signaling (Mallory and Vaucheret, 2006), a pair of natural antisense transcripts (NAT) generating nat-siRNAs in response to salt stress (Borsani *et al.*, 2005), and some chromatin-associated siRNAs in transcriptional gene silencing (Chan *et al.*, 2005; Mallory and Vaucheret, 2006; Matzke and Birchler, 2005).

Research in our laboratory has provided the first example of the regulation of a plant endogenous siRNA in response to a bacterial pathogen, *Pseudomonas syringae*, and its role in plant immunity (Katiyar-Agarwal *et al.*, 2006). By using 454 high-throughput deep-sequencing technology (http://www.454.com, Margulies *et al.*, 2005), we discovered many new endogenous siRNAs and putative miRNAs specifically induced by bacterial and fungal pathogens (Jin *et al.*, unpublished data). Thus, many endogenous small RNAs are regulated by pathogen infection and may play important roles in gene-expression reprogramming in plant defense responses.

In this chapter, we introduce the major experimental approaches, including powerful high-throughput deep-parallel sequencing and Northern hybridization-based technologies, for the discovery and detection of pathogen-regulated small RNAs. We mainly focus on methods of LMW RNA extraction from pathogen-infected tissue and strategies for detecting endogenous small RNAs by Northern blot analysis. Several protocols are compared and described in detail.

2. SEQUENCING-BASED APPROACHES FOR THE DISCOVERY OF PATHOGEN-REGULATED SMALL RNAs

Many miRNAs can be predicted by computational analysis (Jones-Rhoades and Bartel, 2004; Lu *et al.*, 2005; Meyers, 2006); however, these are limited in number. Most of the newly discovered small RNAs are mainly the endogenous siRNAs, which cannot be easily predicted because they are extremely diverse and normally not conserved across species. Therefore, a robust, experimental-based approach is desired for the discovery of endogenous siRNAs as well as novel miRNAs. Small RNAs have characteristic $5'$-phosphate and $3'$-hydroxyl moieties, which are distinct from typical RNA breakdown products that contain a $5'$-hydroxyl group. These small RNAs can be efficiently cloned using the methods developed by Lau *et al.* (2001), Elbashir *et al.* (2001), and Llave *et al.* (2002). The conventional method for discovery of small RNAs involves cloning of small RNAs, reverse transcription, PCR amplification, and sequencing of single cDNA molecules or longer "concatamers" of these molecules (Aravin *et al.*, 2002; Lagos-Quintana *et al.*, 2001; Reinhart *et al.*, 2002). However, only highly expressed miRNAs and siRNAs can be readily detected with these conventional techniques.

Several new high throughput sequencing techniques, including MPSS (Solexa Inc., http://www.solexa.com) (Brenner *et al.*, 2000), 454 sequencing (454 Life Sciences, http://www.454.com) (Margulies *et al.*, 2005), and DNA polymerase colony (Polony) technology (Mitra *et al.*, 2003; Shendure *et al.*, 2005), have emerged. MPSS and 454 sequencing have been demonstrated to be the most robust approaches for small RNA discovery (Girard *et al.*, 2006; Lu *et al.*, 2005, 2006). Instead of cloning the small RNA libraries into plasmid vectors for conventional sequencing, both MPSS and 454 approaches directly amplify and sequence small RNA libraries on microbeads and yield hundreds of thousands of sequence reads simultaneously. The MPSS sequencing reaction is not conventional and proceeds by enzymatic digestion, adapter ligation, amplification, and hybridization to yield about 17 nt of sequence (Brenner *et al.*, 2000; Lu *et al.*, 2005). It requires further analysis to gain the information of the length of the small RNAs. The length limitation of the MPSS sequencing reads also increase the uncertainty for computational analysis of the small RNA loci in the genome. 454 sequencing involves generation and detection of pyrophosphate signal and photons, yielding up to 100 nt of sequence (Margulies *et al.*, 2005). Both technologies include the construction of small RNA libraries, deep-parallel sequencing, and computational analysis, which provide quantitative/semi-quantitative results. Currently, 454 sequencing is more accessible and economical than MPSS; in particular, 454 enables sequencing of the full-length of small RNAs. New sequencing technologies are being developed for deeper coverage, higher sequencing quality, and lower cost. For example, Solexa, Inc. is developing a new technology, sequencing by synthesis (SBS), which uses four proprietary, fluorescently labeled, modified nucleotides to sequence millions of DNA or cDNA fragments in a single run (http://www.solexa.com). In the presence of all four nucleotides, the polymerase is able to select the correct base to incorporate, therefore, leading to high accuracy. SBS amplifies templates on solid surfaces as clusters instead of involving microbeads as in MPSS and 454 sequencing. The features of these three high-throughput parallel-sequencing methods have been summarized in Table 12.1.

To identify small RNAs regulated by plant pathogens, small RNA libraries prepared from pathogen-infected and mock-treated plant material are used for deep parallel sequencing. To ensure meaningful quantitative information recovered from the sequencing result, samples need to be prepared under exactly the same conditions. The small RNA library construction involves resolving the small RNA fraction on a urea-polyacrylamide gel electrophoresis (PAGE) gel and excising 20- to 30-nt small RNAs from the gel. The gel-purified small RNAs are subsequently ligated to 3'- and 5'-RNA adaptors

Table 12.1 Approaches of high-throughput parallel sequencing for small RNA discovery

	MPSS	454	SBS[a]
Reaction	Enzyme digestion and amplification	Emulsified PCR in a water-in-oil mixture	Solid-phase bridge amplification
Signal	Fluorescent signal	Pyrophosphate and photons	Fluorescent signal
Reaction base	Microbeads within a flow cell	Microbeads within a PicoTiterPlate	Solid surface of a flow cell
Length of beads	17 nt	Up to 100 nt	About 30 nt
Websites	www.solexa.com	www.454.com	www.solexa.com

[a] In development.

by use of T4 RNA ligase as described (Elbashir *et al.*, 2001; Lau *et al.*, 2001; Llave *et al.*, 2002). The final ligation products are excised from the gel and then undergo reverse transcription and a low number of PCR cycles for amplification. The same amount of PCR products from both pathogen-treated and mock-treated samples is subjected to deep parallel sequencing. This chapter does not discuss sequence data analysis by bioinformatics approaches. Because MPSS and 454 provide quantitative/semi-quantitative information, small RNAs present at a higher frequency in the infected libraries than in the mock-treated libraries are likely to be inducible by pathogens. Small RNAs absent or present at a lower frequency in the infected libraries than that in the mock-treated libraries are likely to be suppressed by pathogens. Such prediction needs to be validated by further experimental analysis.

3. Hybridization-Based Approaches of Identifying and Validating Pathogen-Inducible Small RNAs

One challenging step for small RNA discovery is to validate and further analyze newly discovered small RNAs from the powerful parallel sequencing. Several approaches for small RNA expression analysis have been developed, including Northern blot analysis, quantitative real-time PCR, and small RNA microarrays (Aravin and Tuschl, 2005; Meyers *et al.*, 2006). Northern blot analysis of individual small RNAs is still the predominant method for small RNA verification. Northern blot analysis is not only reliable and economic, but also provides quantitative information and is accessible for further functional analysis of the small RNAs. The small RNA biogenesis can be analyzed by Northern blot with small RNAs extracted from various silencing pathway mutants. The expression of small RNAs in response to infection with various pathogens can be monitored carefully. For example, with Northern blot analysis, after pathogen infection, many more time points can be examined than with high-throughput parallel sequencing or small RNA microarray analysis, both of which are very expensive. Here, we compare several methods and provide detailed protocols for small RNA extraction from pathogen-infected tissue and hybridization-based Northern blot analysis of small RNAs in plants.

3.1. Extraction of small RNAs from pathogen-infected tissue

The first critical step in small RNA analysis is the isolation of high-quality RNA from plants. This process becomes more challenging when the plants are infected with pathogens. Pathogen-infected plant tissues often display severe symptoms, such as chlorosis, lesions, wilt, or necrosis. These conditions in turn result in accumulation of polysaccharides, secondary metabolites, and degradation products that interfere with RNA quantification, RNA gel processing, and subsequent procedures. To establish the most efficient protocol for extracting total RNA from pathogen-infected leaves, we tested several methods, including the guanidine isothiocyanate (GITC) method, the "hot phenol" method, Trizol (Invitrogen, USA) RNA extraction method, and the sodium chloride-SDS method. Although all methods can yield reasonable total RNAs, the GITC method was the most efficient in providing relatively high-purity and high-yield RNA. Both the Trizol and GITC methods entail only a single step for extraction; however, Trizol extraction gives a considerably lower yield of RNA than the GITC method. Both the "hot phenol" and GITC methods provide high-quality and high-yield total RNA. However, the "hot phenol" method involves handling hot phenol and takes longer for extraction. Therefore, we recommend the GITC method for total RNA extraction from pathogen-infected plant tissue. This method is adapted from the protocol described in Chomczynski and Sacchi (1987). Here we present the stepwise GITC-based RNA extraction protocol.

3.1.1. Total RNA extraction by GITC method
Solutions needed for the GITC method:

- GITC buffer: 4 M guanidine isothiocyanate, 25 mM sodium citrate, and 0.5% N-lauryl sarcosine; add β-mercaptoethanol to a final concentration of 1% prior to use
- 3 M sodium acetate, pH 5.0
- Acid phenol (pH 4.5), chloroform, β-mercaptoethanol

The stepwise GITC RNA extraction protocol:

1. Grind frozen plant tissue in liquid nitrogen to a fine powder with use of a mortar and pestle and add 10 ml of GITC extraction buffer per 1 g of fresh weight of plant material.
2. Incubate the sample at room temperature for 5 min and mix by vortexing. Add 1 ml of 3-M sodium acetate per 1 g of fresh tissue and vortex.
3. Add 10 ml of acid phenol (pH 4.5) per 1 g of fresh tissue and vortex. Incubate at room temperature for 5 min, followed by adding 3.3 ml chloroform. Mix thoroughly and centrifuge the tube at 10,000 rpm for 10 min.
4. Transfer the aqueous phase to a new tube and precipitate the nucleic acids by adding 2 vol of ethanol. Incubate at $-20°$ for at least 1 h.
5. Centrifuge the tube at 12,000 rpm for 15 min. Discard the supernatant and air dry the pellet.

LMW RNA can be separated from high-molecular-weight (HMW) RNA in the resulting total RNA pellet by two generally used methods: polyethylene glycol/NaCl (PEG/NaCl) method or lithium chloride (LiCl) method. Both methods yield a similar amount and quality of LMW RNA. Therefore, we describe both protocols next. We routinely use LiCl for separating LMW RNA.

Protocol for separating LMW RNAs—LiCl method:

1. The total RNA pellet obtained from the earlier RNA extraction process is resuspended in 4 M LiCl at a ratio of 1 ml per 1 g of fresh plant material. In theory, LMW RNA can be dissolved in 4-M LiCl solution, whereas HMW RNA cannot. Disruption of the pellet by pipetting the solution up and down with use of a wide bore tip is recommended to fully dissolve the RNA pellet. Incubating the tubes on ice for 0.5 to 1 h will help increase the yield of LMW RNA. After thorough resuspension, the solution should be milky white.
2. Centrifuge the tube at 13,000 rpm for 5 min.
3. Transfer the supernatant to a new tube.
4. The pellet consists of an HMW fraction of RNA and may be processed further for pure HMW RNA.
5. Repeat steps 1 to 3 to recover more LMW RNA.
6. Precipitate the supernatant with equal volumes of isopropanol. Incubate at $-20°$ for at least 1 h.
7. Centrifuge the tube containing the precipitated LMW RNAs at 13,000 rpm for 25 min. Discard the supernatant.
8. Wash the pellet with 80% ethanol and air-dry.
9. Dissolve the LMW RNA pellet in nuclease-free water.

Protocol for separating LMW RNAs—PEG/NaCl method:

1. The total RNA pellet is dissolved in 500-μl diethyl pyrocarbonate (DEPC)-treated nuclease-free water. Add an equal vol of PEG/NaCl precipitation solution (20% PEG-8000, 2 M NaCl). Mix and keep on ice for at least 30 min.
2. Centrifuge at 13,000 rpm for 15 min. Transfer the supernatant to a new tube.
3. Precipitate small RNAs by adding 2.5 vol of 100% ethanol. Incubate at $-20°$ for at least 3 h.
4. Centrifuge at 13,000 rpm for 15 min. Wash with 80% ethanol.
5. Air-dry the pellet and dissolve in nuclease-free water.

3.2. Electrophoresis and gel transfer of LMW RNAs

LMW RNA isolated from infected plant tissues is resolved on a 17% denaturing PAGE gel (8 M urea-PAGE). Preprocessing the gel at 100 V for 1 h improves the separation of RNA samples. To ensure the detection of low abundant endogenous siRNAs, we recommend loading a large amount of LMW RNA, about 50 to 100 μg or more. To identify small RNAs specifically regulated by pathogens, an uninfected or mock-infected control sample should always be included. To the RNA samples, add an equal volume of gel-loading buffer (95% formamide, 18 mM EDTA, 0.025% SDS, xylene cyanol, and bromophenol blue) and denature the samples at 65° for 10 min, then snap cool on ice for 5 min. Load the RNA samples and conduct the electrophoresis at 200 V for 4 to 6 h depending on the extent of resolution required. For samples infected with pathogens, we process the gel for 6 h to efficiently separate small RNA bands and minimize the running errors due to traces of contaminants present during RNA preparation. The gel is stained with ethidium bromide for 5 min to check the quality of RNA. LMW RNAs are transferred to Hybond N$^+$ (GE Healthcare, USA) membrane and fixed on the membrane in a ultraviolet (UV) cross-linker.

3.3. Hybridization and detection of small RNAs

Several strategies exist for detecting small RNAs in Northern blot analysis. Prior to hybridization, the RNA blot is prehybridized in Perfecthyb solution (Sigma, USA) containing sheared salmon sperm DNA at a final concentration of 100 ug/ml at 38° for at least 2 h.

3.3.1. Detection of small RNAs by use of end-labeled oligonucleotide probe

Northern blot analysis with end-labeled oligonucleotide probes can be used for small RNA detection if the sequence of the small RNA is available, either from small RNA cloning, database searching, or deep, high-throughput sequencing. An oligonucleotide probe complementary to the small RNA sequence can be used to detect the small RNA specifically induced in response to pathogen attack. The following describes the protocol for preparing the 5′ end-labeled oligonucleotide probes:

1. Label 20 pmol of oligonucleotide with 3 μl of 10× T4 polynucleotide kinase buffer, 5μl gamma-P^{32} ATP (6000 Ci/mmol; Perkin Elmer) and 20 U of T4 polynucleotide kinase. The reaction is incubated at 37° for 30 min.
2. Purify the probe with use of a Sephadex G-25 column or commercially available premade columns.
3. Denature the probe at 95° for 5 min and place on ice.
4. Add the probe to the prehybridization solution.
5. Perform the hybridization at 38° for 16 h.
6. Wash the membrane with a series of washing solutions of increasing stringency. Each wash is performed at room temperature for 15 min. The composition of wash solutions are as follows:
   Solution I: 5× SSC, 0.1% SDS
   Solution II: 2× SSC, 0.1% SDS
   Solution III: 1× SSC, 0.1% SDS

Solution IV: $0.1\times$ SSC, 0.5% SDS

In general, washes at room temperature are sufficient to eliminate nonspecific hybridization. In cases of high background effects, higher temperatures such as 38 or 42° may be useful.

7. Expose the blot to a phosphor imager screen or an X-ray film.

We have identified an endogenous siRNA generated from a NAT pair that is specifically induced in plants infected with *P. syringae* pathovar tomato *(Pst)* strain DC3000 carrying the effector *avrRpt2* (Katiyar-Agarwal *et al.*, 2006). This siRNA down-regulates a putative negative regulator of resistance gene *RPS2*-mediated race-specific resistance. Here we describe the analysis of the induction of this small RNA during *Pst* infection time course shown in Fig. 12.1. We used a 5′ end-labeled oligonucleotide probe with the complementary sequence to this small RNA.

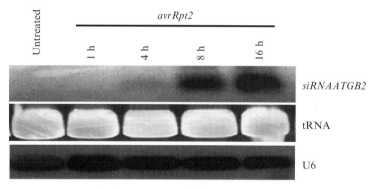

Figure 12.1 Small RNA accumulation in *Arabidopsis* plants in response to infection with *Pseudomonas syringae* carrying the effector *avrRpt2*. The figure shows the time kinetics of accumulation of siRNAATGB2. Small RNA extracted from 4-week-old *Arabidopsis* plants infiltrated with 2×10^7 CFU/ml. Tissue was harvested at different time points as shown on the top. Small RNAs were resolved on 17% urea-PAGE gel and transferred to Hybond N+ membrane. RNA was immobilized on the membrane ultraviolet (UV) cross-linker. Prehybridization was performed in Perfecthyb solution (Sigma, USA) at 38° for at least 2 h. An oligonucleotide complementary to siRNAATGB2 was end-labeled with gamma-P^{32} ATP and T4 polynucleotide kinase. After purification with a Sephadex G-25 column, the probe was denatured at 95° for 5 min and added to the prehybridization solution. Hybridization was performed at 38° for 16 h. Blots were washed as described in the protocol and exposed to the phosphor imager screen. The middle panel shows ethidium bromide–stained tRNA, and the lower panel shows the same blot hybridized with U6 oligonucleotide probe used as a loading control.

3.3.2. Use of a riboprobe to detect small RNAs

Small RNA can also be detected by using a riboprobe in the Northern hybridization. This method is particularly useful when the exact sequence of a small RNA is not available. In many cases, small RNA formation is predicted in the genes or intergenic regions, such as transgene loci, transposable elements, tandem repeats, inverse repeats, or the overlapping regions of natural antisense transcripts, but a small RNA sequence may not be available. Many such examples exist in plant–pathogen interaction studies. Most plant-disease resistance genes are characterized as containing a nucleotide-binding site (NBS) and leucine-rich-repeat (LRR) domain. Many closely related NBS-LRR genes tend to be arranged in clusters in the plant genome, which has the potential to generate small RNAs (Meyers *et al.*, 2005). Furthermore, genome-wide searches in plants and animals have revealed that a large proportion of the eukaryotic genomes are arranged as overlapping NATs in spite of large intergenic regions. More than 1000 pairs of NATs have been identified in the *Arabidopsis* genome. These overlapping antisense transcripts also have the potential to form double-stranded RNAs that could give rise to small RNAs. Our expression analysis indicates that many of these NAT genes are regulated by biotic stress, which has the potential to generate small RNAs to regulate plant defense responses (Jin *et al.*, unpublished data). The formation of small RNAs in any of these circumstances would be interesting to study.

If no small RNA sequences are available within a given region of interest, a hydrolyzed strand-specific riboprobe across this region can be used for Northern blot analysis. For preparing the riboprobe, the fragment is amplified with the forward primer containing the bacteriophage promoter sequence for T7, SP6, or T3 polymerases, or the fragment is cloned in a plasmid vector, such as pGEMT-Easy (Promega), which allows the *in vitro* synthesis of RNA transcripts. A radioisotope-labeled RNA probe is synthesized by using a Riboprobe *in vitro* Transcription System (Promega). The DNA template is removed by treatment with 1 U of RQ1 DNase/μg of DNA at 37° for 20 min. Carbonate hydrolysis of the radiolabeled riboprobe is performed as follows:

1. To the 20-μl riboprobe reaction mix, add 300μl carbonate buffer (120 mM Na$_2$CO$_3$/80 mM NaHCO$_3$). Incubate at 60° for the time calculated by the following equation:

$$T(time) = (L_i - L_f)/KL_iL_f,$$

where L_i is the initial length of probe
L_f is the final length of probe and
$K = 0.11$ kb/min.

2. After hydrolysis, neutralize the reaction with 20 μl 3-M sodium acetate, pH 5.0.

3. Precipitate with 0.1 vol 3-M sodium acetate, pH 5.2, 2-μl glycogen (10 mg/ml), and 2 vol of 100% ethanol. Place the tube at $-20°$ for 30 min.

4. Wash with 80% ethanol, air-dry the pellet, and dissolve in 30 μl nuclease-free water.

5. Denature the probe at 80° for 5 min and add to the prehybridization buffer.

6. Perform the hybridization at 32° for 16 h.

7. Wash and process the blots as previously mentioned.

For detecting low abundant small RNAs, real-time RT-PCR is recommended for maximum sensitivity. An miRNA quantification kit, TaqMan MicroRNA Assays, was developed by Applied Biosystems, Inc. During cDNA synthesis, a universal primer was added to the miRNA for PCR amplification and quantification. This method enables detection of miRNAs with a starting material of only 1 to 10 ng of total RNA.

Finally, the development of small RNA microarray technology makes it possible to assess the expression of many small RNAs simultaneously. In theory, all of the small RNAs identified from high-throughput parallel sequencing can be spotted on the microarray slides for expression analysis. miRNAs and high abundant siRNAs can be easily detected. However, it is a challenge to develop a more sensitive labeling system for detecting many more low abundant endogenous small RNAs.

4. Concluding Remarks

Small RNA-mediated gene silencing has been shown to play an essential role in antiviral defense in both plant and animal systems (Li and Ding, 2005; Voinnet, 2005; Wang *et al.*, 2006). However, these siRNAs are generated from viral RNAs instead of host endogenous RNAs. Defense regulation mediated by endogenous small RNAs has been reported in only a few cases (Katiyar-Agarwal *et al.*, 2006; Lecellier *et al.*, 2005; Navarro *et al.*, 2006). Therefore, the methods described here have great potential to identify novel small RNAs that are induced or repressed by different pathogen infections. High-throughput deep parallel sequencing, coupled with hybridization-based analysis, enables the discovery of many endogenous small RNAs in plant responses to various pathogens. The discovery and analysis of these pathogen-regulated small RNAs will help elucidate the molecular mechanism of plant disease resistance and defense responses and may unravel new components of disease resistance and defense pathways.

REFERENCES

Ambros, V. (2003). MicroRNA pathways in flies and worms: Growth, death, fat, stress, and timing. *Cell* **113,** 673–676.

Aravin, A. A., Lagos-Quintana, M., Yalcin, A., Zavolan, M., Marks, D., Snyder, B., Gaasterland, T., Meyer, J., and Tuschl, T. (2003). The small RNA profile during *Drosophila melanogaster* development. *Developmental Cell* **5,** 337–350.

Aravin, A., and Tuschl, T. (2005). Identification and characterization of small RNAs involved in RNA silencing. *FEBS Lett.* **579,** 5830–5840.

Baulcombe, D. (2005). RNA silencing. *Trends Biochem. Sci.* **30,** 290–293.

Borsani, O., Zhu, J., Verslues, P. E., and Zhu, J. K. (2005). Endogenous siRNAs derived from a pair of natural cis-antisense transcripts regulate salt tolerance in *Arabidopsis. Cell* **123,** 1279–1291.

Brenner, S., Johnson, M., Bridgham, J., Golda, G., Lloyd, D. H., Johnson, D., Luo, S. J., McCurdy, S., Foy, M., Ewan, M., Roth, R., George, D., *et al.* (2000). Gene expression analysis by massively parallel signature sequencing (MPSS) on microbead arrays. *Nat. Biotech.* **18,** 630–634.

Chan, S. W., Henderson, I. R., and Jacobsen, S. E. (2005). Gardening the genome: DNA methylation in *Arabidopsis thaliana. Nat. Rev. Genet.* **6,** 351–360.

Chomczynski, P., and Sacchi, N. (1987). Single-step method of RNA isolation by acid guanidinium thiocyanate-phenol-chloroform extraction. *Anal. Biochem.* **162,** 156–159.

Elbashir, S. M., Lendeckel, W., and Tuschl, T. (2001). RNA interference is mediated by 21-and 22-nucleotide RNAs. *Genes Dev.* **15,** 188–200.

Fujii, H., Chiou, T. J., Lin, S. I., Aung, K., and Zhu, J. K. (2005). A miRNA involved in phosphate-starvation response in *Arabidopsis. Curr. Biol.* **15,** 2038–2043.

Girard, A., Sachidanandam, R., Hannon, G., and Carmell, M. (2006). A germline-specific class of small RNAs binds mammalian Piwi proteins. *Nature* **442,** 199–202.

Henderson, I. R., Zhang, X. Y., Lu, C., Johnson, L., Meyers, B. C., Green, P. J., and Jacobsen, S. E. (2006). Dissecting *Arabidopsis thaliana* DICER function in small RNA processing, gene silencing and DNA methylation patterning. *Nat. Genet.* **38,** 721–725.

Jones-Rhoades, M. W., and Bartel, D. P. (2004). Computational identification of plant MicroRNAs and their targets, including a stress-induced miRNA. *Mol. Cell* **14,** 787–799.

Jones-Rhoades, M. W., Bartel, D., and Bartel, B. (2006). MicroRNAs and their regulatory roles in plants. *Annu. Rev. Plant Biol.* **57,** 19–53.

Katiyar-Agarwal, S., Morgan, R., Dahlbeck, D., Borsani, O., Villegas, A., Zhu, J-K., Staskawicz, B., and Jin, H. (2006). A Pathogen-Inducible endogenous siRNA in plant immunity. *Proc. Natl. Acad. Sci. USA* **103,** 18002–18007.

Lagos-Quintana, M., Rauhut, R., Lendeckel, W., and Tuschl, T. (2001). Identification of novel genes coding for small expressed RNAs. *Science* **294,** 853–858.

Lau, N. C., Lim, L. P., Weinstein, E. G., and Bartel, D. P. (2001). An abundant class of tiny RNAs with probable regulatory roles in *Caenorhabditis elegans. Science* **294,** 858–862.

Li, H. W., and Ding, S. W. (2005). Antiviral silencing in animals. *FEBS Lett.* **579,** 5965–5973.

Llave, C., Kasschau, K. D., Rector, M. A., and Carrington, J. C. (2002). Endogenous and silencing-associated small RNAs in plants. *Plant Cell* **14,** 1605–1619.

Lu, C., Kulkarni, K., Souret, F., Muthuvalliappan, R., Tej, S., Poethig, R., Henderson, I., Jacobsen, S., Wang, W., Green, P. J., and Meyers, B. C. (2006). MicroRNAs and other small RNAs enriched in the *Arabidopsis* RNA-dependent RNA polymerase-2 mutant. *Genome Res.* **16,** 1276–1288.

Lu, C., Tej, S. S., Luo, S., Haudenschild, C. D., Meyers, B. C., and Green, P. J. (2005). Elucidation of the small RNA component of the transcriptome. *Science* **309,** 1567–1569.

Mallory, A. C., and Vaucheret, H. (2006). Functions of microRNAs and related small RNAs in plants. *Nat. Genet.* **38,** S31–S36.

Margulies, M., Egholm, M., Altman, W. E., Attiya, S., Bader, J. S., Bemben, L. A., Berka, J., Braverman, M. S., Chen, Y. J., Chen, Z., Dewell, S. B., Du, L., *et al.* (2005). Genome sequencing in microfabricated high-density picolitre reactors. *Nature* **437,** 376–380.

Matzke, M. A., and Birchler, J. A. (2005). RNAi-mediated pathways in the nucleus. *Nat. Rev. Genet.* **6,** 24–35.

Meyers, B. C., Kaushik, S., and Nandety, R. S. (2005). Evolving disease resistance genes. *Curr. Opin. Plant Biol.* **8,** 129–134.

Meyers, B. C., Souret, F. F., Lu, C., and Green, P. J. (2006). Sweating the small stuff: MicroRNA discovery in plants. *Curr. Opin. Biotechnol.* **17,** 139–146.

Mitra, R. D., Shendure, J., Olejnik, J., Edyta Krzymanska, O., and Church, G. M. (2003). Fluorescent *in situ* sequencing on polymerase colonies. *Anal. Biochem.* **320,** 55–65.

Navarro, L., Dunoyer, P., Jay, F., Arnold, B., Dharmasiri, N., Estelle, M., Voinnet, O., and Jones, J. D. (2006). A plant miRNA contributes to antibacterial resistance by repressing auxin signaling. *Science* **312,** 436–439.

Plasterk, R. H. (2002). RNA silencing: The genome's immune system. *Science* **296,** 1263–1265.

Rajagopalan, R., Vaucheret, H., Trejo, J., and Bareel, D. P. (2006). A diverse and evolutionarily fluid set of microRNAs in *Arabidopsis thaliana. Genes Dev.* **20**(24), 3407–3425.

Reinhart, B. J., Weinstein, E. G., Rhoades, M. W., Bareel, B., and Bareel, D. P. (2002). MicroRNAs in plants. *Genes Dev.* **16**(13), 1616–1626.

Shendure, J., Porreca, G. J., Reppas, N. B., Lin, X., McCutcheon, J. P., Rosenbaum, A. M., Wang, M. D., Zhang, K., Mitra, R. D., and Church, G. M. (2005). Accurate multiplex polony sequencing of an evolved bacterial genome. *Science* **309,** 1728–1732.

Sontheimer, E. J., and Carthew, R. W. (2005). Silence from within: Endogenous siRNAs and miRNAs. *Cell* **122,** 9–12.

Sunkar, R., and Zhu, J. K. (2004). Novel and stress-regulated microRNAs and other small RNAs from *Arabidopsis. Plant Cell* **16,** 2001–2019.

Sunkar, R., Kapoor, A., and Zhu, J. K. (2006). Posttranscriptional induction of two Cu/Zn superoxide dismutase genes in *Arabidopsis* is mediated by downregulation of miR398 and important for oxidative stress tolerance. *Plant Cell* **18,** 2051–2065.

Voinnet, O. (2005). Induction and suppression of RNA silencing: Insights from viral infections. *Nat. Rev. Genet.* **6,** 206–220.

Wang, X. H., Aliyari, R., Li, W. X., Li, H. W., Kim, K., Carthew, R., Atkinson, P., and Ding, S. W. (2006). RNA interference directs innate immunity against viruses in adult *Drosophila. Science* **312,** 452–454.

Waterhouse, P. M., Wang, M. B., and Finnegan, E. J. (2001). Role of short RNAs in gene silencing. *Trends Plant Sci.* **6,** 297–301.

CHAPTER THIRTEEN

PROTOCOLS FOR EXPRESSION AND FUNCTIONAL ANALYSIS OF VIRAL MICRORNAS

Eva Gottwein *and* Bryan R. Cullen

Contents

Abstract

MicroRNAs (miRNAs) are small RNAs of generally 21 to 23 nt that down-regulate target gene expression at the posttranscriptional level. Although miRNAs are endogenously expressed by all animals and plants, numerous DNA viruses have now been identified that also encode miRNAs, presumably to down-modulate protein expression from viral and host transcripts. Although this has been shown in some cases, the function of the majority of viral miRNAs remains unclear. The herpesviruses stand out by making extensive use of miRNA expression during long-term latent infection or lytic replication. Because viral miRNAs are only present in the context of viral infection and can therefore be considered "exogenous," their expression in uninfected cells of the appropriate cell type is a valuable tool to assess their function. Techniques to achieve and validate the expression of viral miRNAs are described in this review.

Center for Virology and Department of Molecular Genetics and Microbiology, Duke University Medical Center, Durham, North Carolina

Methods in Enzymology, Volume 427
ISSN 0076-6879, DOI: 10.1016/S0076-6879(07)27013-2

1. INTRODUCTION

miRNAs are noncoding RNAs, most commonly 21 to 23 nucleotides (nts) in length, that posttranscriptionally down–regulate protein expression from mRNAs containing sequences with varying degrees of complementarity to the miRNA (Bartel, 2004). Vertebrate genomes contain several hundred miRNA genes (http://microrna.sanger.ac.uk/sequences), and more recently, miRNAs have also been identified in several animal DNA viruses (Cullen, 2006c). The herpesviruses, a group of DNA viruses characterized by their large (~150 kb) genome and nuclear replication, are unique in making extensive use of miRNAs. The human pathogenic γ-herpesviruses, Epstein-Barr Virus (EBV) and Kaposi's sarcoma–associated herpesvirus (KSHV), express 23 and 12 miRNAs, respectively (Cai *et al.*, 2005, 2006; Grundhoff *et al.*, 2006; Pfeffer *et al.*, 2004, 2005; Samols *et al.*, 2005). miRNAs have also been cloned from the γ-herpesvirus rhesus lymphocryptovirus (rLCV, 16 miRNAs) (Cai *et al.*, 2006), which belongs to the same genus as EBV, and mouse herpesvirus 68 (MHV68, 9 miRNAs) (Pfeffer *et al.*, 2005), which is a rhadinovirus like KSHV. Members of the other two herpesvirus families, α-herpesviruses and β-herpesviruses, have also been demonstrated to express miRNAs. The herpes simplex virus 1 (HSV-1) latency-associated transcript (LAT) was reported to be processed into at least one miRNA (Gupta *et al.*, 2006) and one miRNA gene upstream of the LAT transcription start site was predicted and validated (Cui *et al.*, 2006). The avian α-herpesvirus Marek's disease virus encodes eight miRNAs, three of which also map to its LAT gene (Burnside *et al.*, 2006). The β-herpesvirus human cytomegalovirus (HCMV) expresses 11 miRNAs from multiple loci (Dunn *et al.*, 2005; Grey *et al.*, 2005; Pfeffer *et al.*, 2005). Apart from the herpesviruses, the polyomavirus SV40 expresses a single miRNA late in its life cycle (Sullivan *et al.*, 2005), and the adenovirus VAI noncoding RNA is processed by Dicer to yield a RISC–associated small RNA (Sano *et al.*, 2006). Although cellular miRNAs are generally highly conserved, most of the known viral miRNAs share no homology with either cellular miRNAs or other viral miRNAs. Exceptions are seven miRNAs conserved between EBV and the related rLCV (Cai *et al.*, 2006), as well as some HCMV and chimpanzee CMV miRNAs (Grey *et al.*, 2005).

The design of effective miRNA expression cassettes is facilitated by a detailed understanding of miRNA biogenesis. Viral and cellular primary miRNA precursors (pri-miRNAs) are typically transcribed by RNA PolII (Cai *et al.*, 2004; Lee *et al.*, 2004), the only known exception being the MHV68 miRNAs, which are expressed as tRNA fusions and transcribed from PolIII promoters (Cai *et al.*, 2004; Pfeffer *et al.*, 2005). The miRNA sequence is contained within one arm of a stem-loop structure within the primary transcript and liberated by two consecutive cleavage events

(Cullen, 2004). In the nucleus, the microprocessor complex containing the RNaseIII enzyme Drosha and the dsRNA binding protein DGCR8 liberates the apical part of the stem-loop structure, the pre-miRNA intermediate (Denli *et al.*, 2004; Gregory *et al.*, 2004; Lee *et al.*, 2003b). Pre-miRNAs are ~60- to 70-nt long and characterized by a 2-nt 3′ OH overhang. Although pre-miRNAs hairpins are most commonly found in introns (Kim and Nam, 2006; Rodriguez *et al.*, 2004), they can also be contained in noncoding RNAs, and, in rare cases, also in coding sequences and 3′UTRs (Cai *et al.*, 2005; Grundhoff *et al.*, 2006; Kim and Nam, 2006). In the latter case, the coding potential of the relevant mRNA will be impaired by Drosha cleavage of the transcript.

The pre-miRNA is recognized by Exportin 5 and transported to the cytoplasm (Lund *et al.*, 2004; Yi *et al.*, 2003), where it is cleaved by the RNaseIII enzyme Dicer to give an imperfect RNA duplex with a 2-nt 3′ OH overhang at each end (Bernstein *et al.*, 2001; Grishok *et al.*, 2001; Hutvagner *et al.*, 2001; Ketting *et al.*, 2001; Knight and Bass, 2001). The strand with the weaker 5′ base pairing generally represents the mature miRNA and is assembled into the RNA-induced silencing complex (RISC) (Khvorova *et al.*, 2003; Schwarz *et al.*, 2003). RISC-bound miRNAs can guide the cleavage of mRNAs when a perfectly complementary target site is present (Hutvagner and Zamore, 2002; Yekta *et al.*, 2004; Zeng *et al.*, 2003). In the case of animal miRNAs, however, target mRNA cleavage seems to be the exception (Yekta *et al.*, 2004), and typically, miRNA:mRNA matches exhibit more limited complementarity, which commonly leads to the inhibition of target mRNA translation (Doench *et al.*, 2003; Olsen and Ambros, 1999; Zeng *et al.*, 2003). It is believed that cellular miRNAs have multiple to several hundred targets, and miRNAs also appear to act coordinately (Krek *et al.*, 2005; Lewis *et al.*, 2005).

In most cases, the functions of viral miRNAs are not known. The HSV-1 LAT-derived miRNA was proposed to be the mediator of the antiapoptotic properties of LAT by targeting TGF-β and Smad3 mRNAs (Gupta *et al.*, 2006). The SV40 miRNA (sv40-mir-S1) is perfectly complementary to the early viral mRNAs and mediates their degradation, thereby resulting in the down-regulation of T antigen expression during the later stages of the replication cycle (Sullivan *et al.*, 2005). Because T antigen is a target for recognition by cytotoxic T cells (CTLs), its down-regulation leads to the escape of virus-infected cells from CTL-mediated killing (Sullivan *et al.*, 2005). One EBV miRNA and several HCMV miRNAs are perfectly complementary to mRNAs originating from the DNA strand opposing the miRNA locus (Grey *et al.*, 2005; Pfeffer *et al.*, 2005), but the significance of the potential down-regulation of these genes to the viral life cycle is not understood. Thus, the majority of viral miRNAs are likely to be specific for host cell mRNAs.

Several strategies can be used to assess the function of viral miRNAs. In general, the miRNA(s) of interest can be overexpressed in noninfected cells or one can interfere with miRNA function in the context of virally infected cells. In this chapter, techniques are described to achieve the expression of physiological levels of viral miRNAs and to monitor the activity of the expressed miRNAs. It is anticipated that techniques to express viral miRNAs and to assess their functionality will greatly facilitate future functional studies.

2. miRNA Expression Cassettes

Ways to achieve the introduction of a viral miRNA sequences into cells include: (1) the transient transfection of synthetic siRNA duplexes with the desired sequence (Elbashir et al., 2001); or (2) the transient or stable delivery of RNA precursors by transfection or viral transduction. Plasmid or viral vector-based delivery have the advantage of lower cost and provide the option to achieve stable expression of the viral miRNA. As most herpesviral miRNAs are stably expressed during viral latency (Cai et al., 2005, 2006; Pfeffer et al., 2004, 2005), the latter approach is closest to their viral expression pattern. Furthermore, viral transduction allows the expression of miRNAs in hard-to-transfect cell types, including primary cells, and the stable expression of miRNAs may reveal miRNA targets that have half-lives longer than the transient expression of an miRNA lasts.

miRNA expression cassettes can be composed of short hairpin RNAs (shRNAs) (Brummelkamp et al., 2002; Paddison et al., 2002), which mature by Dicer cleavage, or longer pri-miRNA–like precursors, from which mature miRNAs are produced using the complete cellular miRNA processing pathway (Chen et al., 2004; Zeng et al., 2002). Which type of expression cassette to choose will depend on whether transient or stable introduction of the miRNA is to be achieved as well as the choice of cell line and delivery system (see following). In our experience, the results achieved with each strategy depend on the miRNA sequence, and several cassettes may have to be cloned to optimize expression. The transcription of shRNAs requires a PolIII promoter, whereas pri-miRNAs can be transcribed from either a PolII or PolIII promoter. PolII promoters have the advantage that expression can be regulated (e.g., using a Tet-inducible promoter) and that cell type–specific promoters can be chosen. Moreover, there is evidence that PolII promoter-driven cassettes lead to more efficient entry of miRNAs into RISC (Chang et al., 2006). Although shRNAs will typically not result in efficient expression when present at low copy number, efficient expression of PolII-driven miRNA cassettes, even at a single copy per cell, has been reported (Stegmeier et al., 2005).

The construction and expression of shRNAs has been reviewed previously (Cullen, 2006b). Researchers have made use of pri-miRNA expression cassettes in which (1) sequences derived from natural pri-miRNAs are used (Chen et al., 2004; Gottwein et al., 2006) or (2) the mature miRNA sequence is embedded into a heterologous pri-miRNA, most commonly that of human miR-30 (Chang *et al.*, 2006; Zeng *et al.*, 2002). Both expression cassettes work under the control of either a PolII or a PolIII promoter. When a PolIII promoter is used, oligo(T) stretches have to be avoided, because this leads to the premature termination of transcription. If miRNAs are to be expressed from viral genomic sequences, using ~250 bp or more from the virus genome with the pre-miRNA sequence at a central position generally works well. It is preferable to express miRNAs from their natural sequence context because this ensures miRNAs are processed correctly and artifacts that can arise due to RNA secondary structure (e.g., the interferon response) as well as the inappropriate expression of the passenger strand are avoided. We observed, however, that some viral miRNAs cannot be readily expressed using their 250-bp genomic regions (not shown). The reason for this is currently unclear, but might involve misfolding of the primary transcript in a way that abolishes processing by Drosha.

If miRNA sequences are to be expressed using a heterologous pri-miRNA stem-loop, such as miR-30, the mature miRNA sequence of this precursor is exchanged for the miRNA in question, and base pairing of the stem is adjusted to preserve the integrity of the stem-loop (Zeng *et al.*, 2002). This strategy has been reviewed in detail elsewhere (Zeng *et al.*, 2005), and there are a few important points to consider when designing an artificial pri-miRNA expression cassette: (1) Base-pairing at the 5′ of the mature miRNA sequence should be weaker than at its 3′-end to avoid expression of the passenger strand. In any case, it should be controlled whether the passenger strand is expressed in cells, because this sequence likely down-regulates at least a few mRNAs that, by chance, bear regions complementary to the passenger strand. (2) Most natural miRNAs contain bulges in their stem, and this may help to avoid triggering a cellular interferon response to dsRNA (Cullen, 2006a). We routinely introduce bulges both into miR-30–based miRNA precursors as well as into shRNAs and have reported excellent expression using these cassettes (Zeng *et al.*, 2005).

3. Vector Systems for Stable Delivery

Many different plasmids and viral vectors that allow the transient or stable expression of pri-miRNA sequences under the control of PolII or PolIII promoters have been described (Amarzguioui *et al.*, 2005; Chang *et al.*, 2006; Dickins *et al.*, 2005; Silva *et al.*, 2005; Stegmeier *et al.*, 2005; Zeng *et al.*, 2005). The viral vector of choice will depend mainly on the cell type that is to be transduced and the miRNA expression cassette (see earlier). Vectors that allow inducible expression of miRNAs or the coupling of miRNA expression with fluorescent or antibiotic marker gene expression have also been described (Dickins *et al.*, 2005; Stegmeier *et al.*, 2005).

For the purpose of this review, we will use the self-inactivating lentiviral vector pNL-SIN-CMV-BLR as an example (Lee *et al.*, 2003a). To achieve expression of the KSHV miRNA miR-K1, 250 bp of the KSHV genome (nt 121762-122003) encompassing pre-miR-K1 at a central position were placed in the 3'UTR of the Blasticidin S deaminase (BLR) gene (using ClaI and XbaI restriction sites) (Fig. 13.1A). Although this vector has previously been used to overexpress shRNAs, driven from the H1 PolIII promoter in the antisense orientation (Lee *et al.*, 2003a), it should be noted that we have here inserted the pri-miRNA fragment in the sense orientation into the 3'UTR of BLR. Importantly, processing of this pri-miRNA cassette by Drosha is not efficient enough to abolish BLR synthesis, and therefore, infected cells can be selected using Blasticidin S. The same approach can be used with any other PolII-driven miRNA cassette. Once an expression vector has been cloned, it is advisable to test miRNA expression from this vector after transient transfection. In our hands, calcium phosphate transfection of 293T cells (e.g., 2 μg lentiviral vector/6 well) works well for this purpose. Forty-eight to seventy-two hours after transfection, cells are harvested by lysis in TRIzol (Invitrogen) and total RNA is prepared. In general, mature miRNA expression can be readily confirmed using

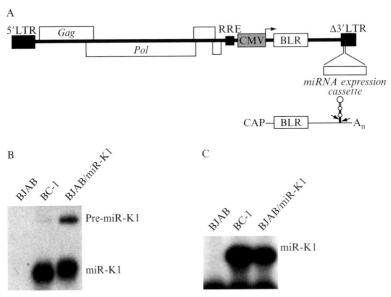

Figure 13.1 Viral microRNA (miRNA) expression strategy. (A) Schematic of the self-inactivating lentiviral vector pNL-SIN-CMV-BLR. The miRNA expression cassette was placed into the 3'UTR of the Blasticidin S resistance gene (BLR). The transcript derived from the integrated vector is shown at the bottom of the panel. (B) Northern analysis of miR-K1 expression in the negative control cell line BJAB, in latently KSHV-infected BC-1 cells and in cells stably transduced with pNL-SIN-CMV-BLR/miR-K1 (BJAB/miR-K1). (C) Primer extension analysis of miR-K1 expression of the samples described in B.

Northern blotting or primer extension. Northern blotting allows the detection of the pre-miRNA regardless of which arm of the pre-miRNA the miRNA derives from. It is important to include both a positive control (virally infected cells, if available) and a negative control (uninfected cells). The reader is referred to a previous review for detailed protocols (Zeng *et al.*, 2005).

Once miRNA expression from the vector has been confirmed in a transient setting, the vector as well as control vectors (e.g., empty vector or a vector expressing a nontargeting miRNA) are packaged and used to infect target cell lines. Before stable cell lines are generated, a Blasticidin S killing assay using the cell line that will be transduced should be conducted (unless the relevant Blasticidin S concentration is known from the literature or previous experiments). For this, cells should be placed under different concentrations of Blasticidin S (typically 1 to 10 μg/ml). The lowest concentration of Blasticidin S that results in killing of all cells within 2 to 4 days should be used for the selection.

4. PROTOCOL: GENERATION OF MIRNA-EXPRESSING CELL LINES USING PNL-SIN-CMV-BLR–BASED MIRNA EXPRESSION VECTORS

Day 1:

1. Plate 293T cells at a confluency of ~60%. Prepare one 10-cm dish for each pNL-SIN-CMV-BLR vector. It is helpful to also include one transfection using the BLR-negative pNL-SIN-CMV-GFP lentiviral vector as a control for transduction efficiency and to monitor killing induced by Blasticidin S during selection (see following, days 6 to 12).

Day 2:

2. Transfect the 293T cells plated on day 1 using an established transfection method. For each 10-cm dish, use 9 μg lentiviral vector, 0.5 μg each of pcRev and pcTat, and 0.25 μg pHITG (an expression plasmid for VSV-G). Very good titers (~10^6 i.u./ml) are typically achieved by calcium phosphate-mediated transfection.
3. Remove the culture medium 4 to 12 h after transfection and add 5 to 12 ml fresh medium per 10-cm dish.

Day 3:

4. Prepare the cells that will be transduced the next day:
   a. In the case of adherent cells, plate one 10-cm dish at ~60% confluency for each transduction.
   b. In the case of suspension cells, split the cell line that will be used for transduction to ~0.5×10^6 cells/ml.

Day 4:

5. Harvest virus-containing cell culture supernatants (~48 h after transfection) and pass through filter (450 nm pore size). Filtering the virus stocks ensures that no cells are carried over from the transfection. Adjust each virus stock to 5 μg/ml polybrene.
6. Infect the cells prepared on day 3:
    a. In the case of adherent cells, remove culture medium and replace with virus-containing medium.
    b. In the case of suspension cells, collect cells by centrifugation and concentrate to ~5 × 10^6/ml. Use 0.5 ml of this concentrated cell suspension per 5 ml of virus in a 10-cm dish.

Day 5:

7. Change culture medium 8 to 24 hours after transduction. For this, collect suspension cells by centrifugation and resuspend in fresh culture medium.

Day 6:

8. Place cells under selection with the predetermined concentration of Blasticidin S. For this, remove the culture supernatant and add medium containing the predetermined concentration of Blasticidin S (see earlier).

Days 6 to 12:

9. Monitor the cells for killing. If close to 100% of the cells are infected, no cells will die and the cultures will be growing as before. If less than 100% cells are infected, uninfected cells will be lost upon selection. Cells should be passaged like the parental cell line (using media containing Blasticidin S). Selection can be considered to be complete when all pNL-SIN-CMV-GFP transduced cells are dead. Cells should be frozen at an early passage and while Blasticidin S is still in the culture medium. Blasticidin S can be removed from the system once selection is complete.

In the example shown in Fig. 13.1, the virus generated by transfection of pNL-SIN-CMV-BLR/miR-K1 was used to infect the KSHV and EBV-negative B cell line BJAB. Forty-eight hours after transduction, cells were placed under selection with Blasticidin S (10 μg/ml). The cell lines generated in this way were analyzed for expression of miR-K1 by Northern and primer extension analyses (see Fig. 13.1B and C) as well as indicator assays (see following). The Northern and primer extension analyses show that the expression of mature miR-K1 in this cell line is very close to its expression in the latently KSHV-infected cell line BC-1. However, pre-miR-K1 accumulated to higher levels in BJAB/miR-K1 cells. This effect has also been observed using other vector systems and may reflect differences in the efficiency of miRNA processing between the BJAB and BC-1 cell lines (not shown).

5. INDICATOR ASSAYS ESTABLISH miRNA ACTIVITY

Indicator assays allow an investigator to demonstrate the biological activity of a specific miRNA by placing sequences with perfect complementarity to the miRNA into the 3′UTR of an indicator gene, such as luciferase. We have adapted the use of indicator assays to hard-to-transfect cell lines by delivery of the indicator using lentiviral transduction rather than transfection (Gottwein *et al.*, 2006). For this purpose, we adapted the previously described HIV-1–based self-inactivating retroviral vector pNL-SIN-CMV (Lee *et al.*, 2003a) to express firefly luciferase (FLuc) or renilla luciferase (RLuc) under the control of the CMV promoter (Fig. 13.2A). Two target sites perfectly complementary to the miRNA in question, in this case miR-K1, were inserted into the 3′UTR of RLuc using synthetic oligonucleotides following the general design shown in Fig. 13.2, panel B. Upon annealing, these oligonucleotides yield ends suitable for ligation into the vector ClaI and XbaI sites. The annealed oligonucleotides also contain two target sites with perfect complementarity to miR-K1 separated by a short spacer sequence. By using the sequences shown, the upstream ClaI site is lost upon ligation and a new ClaI site just downstream of the target sites is introduced. This allows the insertion of more annealed oligonucleotides using the same strategy. If this is not desired, the downstream ClaI site can be omitted from the oligonucleotides used.

6. PROTOCOL: PREPARATION OF VIRUS MIXES AND INDICATOR ASSAY

Day 1:

10. Plate 293T cells at a confluency of ∼60%. Prepare 1 well of a 6-well plate for each RLuc vector used and the same total number of wells for transfection with the FLuc control vector.

Day 2:

11. Transfect the 293T cells (plated on day 1) using an established transfection method. For each 6 well, use 1.8 μg lentiviral vector, 0.1 μg each of pcRev and pcTat, and 0.05 μg pHITG. It is helpful to prepare a mastermix of the packaging constructs. Very good titers (∼10^6 i.u./ml) are typically achieved by calcium phosphate–mediated transfection.
12. Remove the culture medium 4 to 12 h after transfection and replace with 1 to 3 ml fresh medium per well of a 6-well dish.

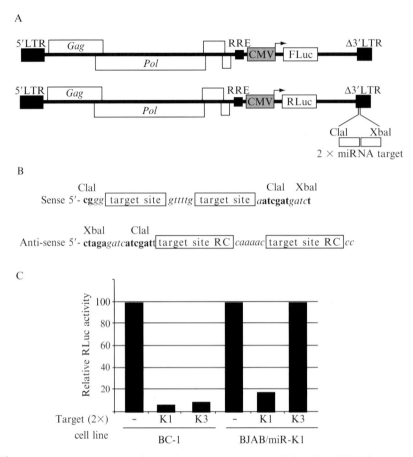

Figure 13.2 Indicator assay for viral microRNA (miRNA) function. (A) Schematic of the structure of the lentiviral RLuc indicator vector and the Fluc control vector. (B) Oligonucleotide design for cloning miRNA target sequences into pNL-SIN-CMV-RLuc. Restriction sites are indicated. RC: reverse complement. (C) BJAB, BC-1, and BJAB cells stably transduced with pNL-SIN-CMV-BLR/miR-K1 (BJAB/miR-K1) were transduced with a mixture of the Fluc control virus and the RLuc indicator viruses carrying no target sites or target sites for miR-K1 or miR-K3. Dual luciferase assays were performed 24 h later. RLuc activities were normalized to the observed FLuc activity. Normalized RLuc activities in KSHV-positive BC-1 and miR-K1-positive BJAB/miR-K1 cells are shown relative to those obtained in KSHV-negative BJAB cells. Values from cells transduced with the parental RLuc-expressing vector were set at 100%.

Day 3:

13. Prepare the cells that will be transduced the next day.
   a. In the case of adherent cells, plate 1 well of a 6-well plate or 24-well plate for each transduction (60% confluent).

b. In the case of suspension cells, split the cell lines that will be used for transduction to ~0.5 × 10⁶ cells/ml.

Day 4:

14. Harvest virus-containing cell culture supernatants (~48 h after transfection) and pass through filter (450 nm pore size). Filtering the virus stocks ensures that no cells are carried over from the transfection.
15. Mix each RLuc virus stock with an equal volume of Fluc control virus and adjust mixtures to 5 μg/ml polybrene.
16. Use 1 ml of each virus mix to infect ~0.5 × 10⁶ suspension cells in a well of a 6-well plate for
    a. a control cell line not expressing the miRNA in question
    b. a cell line expressing the relevant miRNA

If adherent cells are used, remove the medium from the cells plated on day 3 and add 1 ml of virus mix per well of a 6-well plate. We also achieved excellent results by infecting cells in 24-well plates (in this case, 0.2 of the cells and the virus is sufficient). If cells are harvested within 24 h after transduction, it is not necessary to replace the culture medium.

Day 5:

17. Harvest the cells (~24 h after transduction):

    a. In the case of adherent cells: remove culture media, rinse cells with PBS and add 500 μl passive lysis buffer to each 6-well (100 μl to each 24-well).
    b. In the case of suspension cells: collect cells by centrifugation (400×g, 2 min, tabletop centrifuge), remove medium, wash with PBS, collect cells by centrifugation, remove PBS, and lyse cell pellet in 100 to 500 μl 1× passive lysis buffer (Promega).
18. Perform dual luciferase assays (e.g., using the Dual-Luciferase Reporter Assay System [Promega]). Alternatively, the lysates can be stored at −20° for later analysis.
19. Data analysis:

    a. Calculate the ratio of Rluc to Fluc activities for each sample.
    b. Normalize the values obtained for indicator constructs containing target sites to the values obtained with the RLuc control vector, which are set at 100%.
    c. Then, normalize the values obtained from the miRNA-expressing samples to those obtained for the negative control cell line, which are set at 100% for each target site construct.

In the example shown in Fig. 13.2C, mixtures of Fluc virus stock with RLuc virus without miRNA target sites and RLuc viruses bearing target sites specific for the KSHV miRNAs miR-K1 or miR-K3 were used to infect

(1) BJAB cells, which do not express either miRNA and serve as negative control cells; (2) the latently KSHV-infected B cell line BC-1, which naturally expresses both miR-K1 and miR-K3; and (3) the BJAB/miR-K1 cell line generated as described previously. Infections were performed using 5×10^5 cells and 1 ml of virus stock. Twenty-four hours after transduction, cells were collected by centrifugation, the supernatant medium was aspirated, cells were lysed, and dual luciferase assays were carried out using the Dual-Luciferase Reporter Assay System (Promega). The ratio of Rluc to Fluc activities was calculated for each sample and, for each cell line, the values obtained for indicator constructs containing target sites (i.e., for miR-K1 or miR-K3) were normalized to the values obtained with the RLuc control vector, which were set at 100%. The resulting values obtained for BC-1 and BJAB/miR-K1 cells were normalized to those obtained from the negative control cell line BJAB, which were set at 100% for each target site construct. As shown in Fig. 13.2C, in BC-1 cells, RLuc expression from vectors carrying target sites for either miR-K1 or miR-K3 was reduced by approximately 10-fold. This is in good agreement with previously published data (Gottwein *et al.*, 2006) and reflects the activity of endogenous miR-K1 and miR-K3 in the latently KSHV-infected BC-1 cells. In the case of the BJAB/miR-K1 cells, RLuc expression from the vector carrying two target sites complementary to miR-K1 was reduced by more than 5-fold, while expression from the vector carrying target sites complementary to miR-K3 was essentially unaffected. Therefore, we conclude that the levels of miR-K1 activity in the BJAB/miR-K1 cell line generated as described earlier are very close to those observed in KSHV-infected B cells.

7. Concluding Remarks

In this chapter, we have described strategies to generate viral miRNA expression cassettes, lentiviral miRNA expression vectors as well as the use of lentiviral indicator vectors to assess the activity of the ectopically expressed miRNAs. It is anticipated that the artificial expression of physiological levels of viral miRNAs in the appropriate cell types will lead to fundamental insights into the function of these viral miRNAs in the viral life cycle.

ACKNOWLEDGMENTS

E. Gottwein is a recipient of a Feodor Lynen fellowship from the Alexander von Humboldt foundation. This work was funded by NIH grant R01 AI067968.

REFERENCES

Amarzguioui, M., Rossi, J. J., and Kim, D. (2005). Approaches for chemically synthesized siRNA and vector-mediated RNAi. *FEBS Lett.* **579**, 5974–5981.

Bartel, D. P. (2004). MicroRNAs: Genomics, biogenesis, mechanism, and function. *Cell* **116,** 281–297.

Bernstein, E., Caudy, A. A., Hammond, S. M., and Hannon, G. J. (2001). Role for a bidentate ribonuclease in the initiation step of RNA interference. *Nature* **409,** 363–366.

Brummelkamp, T. R., Bernards, R., and Agami, R. (2002). A system for stable expression of short interfering RNAs in mammalian cells. *Science* **296,** 550–553.

Burnside, J., Bernberg, E., Anderson, A., Lu, C., Meyers, B. C., Green, P. J., Jain, N., Isaacs, G., and Morgan, R. W. (2006). Marek's disease virus encodes MicroRNAs that map to meq and the latency-associated transcript. *J. Virol.* **80,** 8778–8786.

Cai, X., Hagedorn, C. H., and Cullen, B. R. (2004). Human microRNAs are processed from capped, polyadenylated transcripts that can also function as mRNAs. *RNA* **10,** 1957–1966.

Cai, X., Lu, S., Zhang, Z., Gonzalez, C. M., Damania, B., and Cullen, B. R. (2005). Kaposi's sarcoma-associated herpesvirus expresses an array of viral microRNAs in latently infected cells. *Proc. Natl. Acad. Sci. USA* **102,** 5570–5575.

Cai, X., Schafer, A., Lu, S., Bilello, J. P., Desrosiers, R. C., Edwards, R., Raab-Traub, N., and Cullen, B. R. (2006). Epstein-Barr virus microRNAs are evolutionarily conserved and differentially expressed. *PLoS Pathog.* **2,** e23.

Chang, K., Elledge, S. J., and Hannon, G. J. (2006). Lessons from nature: MicroRNA-based shRNA libraries. *Nat. Methods* **3,** 707–714.

Chen, C. Z., Li, L., Lodish, H. F., and Bartel, D. P. (2004). MicroRNAs modulate hematopoietic lineage differentiation. *Science* **303,** 83–86.

Cui, C., Griffiths, A., Li, G., Silva, L. M., Kramer, M. F., Gaasterland, T., Wang, X. J., and Coen, D. M. (2006). Prediction and identification of herpes simplex virus 1-encoded microRNAs. *J. Virol.* **80,** 5499–5508.

Cullen, B. R. (2004). Transcription and processing of human microRNA precursors. *Mol. Cell* **16,** 861–865.

Cullen, B. R. (2006a). Enhancing and confirming the specificity of RNAi experiments. *Nat. Methods* **3,** 677–681.

Cullen, B. R. (2006b). Induction of stable RNA interference in mammalian cells. *Gene Ther.* **13,** 503–508.

Cullen, B. R. (2006c). Viruses and microRNAs. *Nat. Genet.* **38**(Suppl.), S25–S30.

Denli, A. M., Tops, B. B., Plasterk, R. H., Ketting, R. F., and Hannon, G. J. (2004). Processing of primary microRNAs by the Microprocessor complex. *Nature* **432,** 231–235.

Dickins, R. A., Hemann, M. T., Zilfou, J. T., Simpson, D. R., Ibarra, I., Hannon, G. J., and Lowe, S. W. (2005). Probing tumor phenotypes using stable and regulated synthetic microRNA precursors. *Nat. Genet.* **37,** 1289–1295.

Doench, J. G., Petersen, C. P., and Sharp, P. A. (2003). siRNAs can function as miRNAs. *Genes Dev.* **17,** 438–442.

Dunn, W., Trang, P., Zhong, Q., Yang, E., van Belle, C., and Liu, F. (2005). Human cytomegalovirus expresses novel microRNAs during productive viral infection. *Cell Microbiol.* **7,** 1684–1695.

Elbashir, S. M., Harborth, J., Lendeckel, W., Yalcin, A., Weber, K., and Tuschl, T. (2001). Duplexes of 21-nucleotide RNAs mediate RNA interference in cultured mammalian cells. *Nature* **411,** 494–498.

Gottwein, E., Cai, X., and Cullen, B. R. (2006). A novel assay for viral microRNA function identifies a single nucleotide polymorphism that affects Drosha processing. *J. Virol.* **80,** 5321–5326.

Gregory, R. I., Yan, K. P., Amuthan, G., Chendrimada, T., Doratotaj, B., Cooch, N., and Shiekhattar, R. (2004). The Microprocessor complex mediates the genesis of micro-RNAs. *Nature* **432,** 235–240.

Grey, F., Antoniewicz, A., Allen, E., Saugstad, J., McShea, A., Carrington, J. C., and Nelson, J. (2005). Identification and characterization of human cytomegalovirus-encoded microRNAs. *J. Virol.* **79**, 12095–12099.

Grishok, A., Pasquinelli, A. E., Conte, D., Li, N., Parrish, S., Ha, I., Baillie, D. L., Fire, A., Ruvkun, G., and Mello, C. C. (2001). Genes and mechanisms related to RNA interference regulate expression of the small temporal RNAs that control *C. elegans* developmental timing. *Cell* **106**, 23–34.

Grundhoff, A., Sullivan, C. S., and Ganem, D. (2006). A combined computational and microarray-based approach identifies novel microRNAs encoded by human gamma-herpesviruses. *RNA* **12**, 733–750.

Gupta, A., Gartner, J. J., Scthupathy, P., Hatzigeorgiou, A. G., and Fraser, N. W. (2006). Anti-apoptotic function of a microRNA encoded by the HSV-1 latency-associated transcript. *Nature* **442**, 82–85.

Hutvagner, G., and Zamore, P. D. (2002). A microRNA in a multiple-turnover RNAi enzyme complex. *Science* **297**, 2056–2060.

Hutvagner, G., McLachlan, J., Pasquinelli, A. E., Balint, E., Tuschl, T., and Zamore, P. D. (2001). A cellular function for the RNA-interference enzyme Dicer in the maturation of the let-7 small temporal RNA. *Science* **293**, 834–838.

Ketting, R. F., Fischer, S. E., Bernstein, E., Sijen, T., Hannon, G. J., and Plasterk, R. H. (2001). Dicer functions in RNA interference and in synthesis of small RNA involved in developmental timing in *C. elegans*. *Genes Dev.* **15**, 2654–2659.

Khvorova, A., Reynolds, A., and Jayasena, S. D. (2003). Functional siRNAs and miRNAs exhibit strand bias. *Cell* **115**, 209–216.

Kim, V. N., and Nam, J. W. (2006). Genomics of microRNA. *Trends Genet.* **22**, 165–173.

Knight, S. W., and Bass, B. L. (2001). A role for the RNase III enzyme DCR-1 in RNA interference and germ line development in *Caenorhabditis elegans*. *Science* **293**, 2269–2271.

Krek, A., Grun, D., Poy, M. N., Wolf, R., Rosenberg, L., Epstein, E. J., MacMenamin, P., da Piedade, I., Gunsalus, K. C., Stoffel, M., and Rajewsky, N. (2005). Combinatorial microRNA target predictions. *Nat. Genet.* **37**, 495–500.

Lee, M. T., Coburn, G. A., McClure, M. O., and Cullen, B. R. (2003a). Inhibition of human immunodeficiency virus type 1 replication in primary macrophages by using Tat- or CCR5-specific small interfering RNAs expressed from a lentivirus vector. *J. Virol.* **77**, 11964–11972.

Lee, Y., Ahn, C., Han, J., Choi, H., Kim, J., Yim, J., Lee, J., Provost, P., Radmark, O., Kim, S., and Kim, V. N. (2003b). The nuclear RNase III Drosha initiates microRNA processing. *Nature* **425**, 415–419.

Lee, Y., Kim, M., Han, J., Yeom, K. H., Lee, S., Baek, S. H., and Kim, V. N. (2004). MicroRNA genes are transcribed by RNA polymerase II. *EMBO J.* **23**, 4051–4060.

Lewis, B. P., Burge, C. B., and Bartel, D. P. (2005). Conserved seed pairing, often flanked by adenosines, indicates that thousands of human genes are microRNA targets. *Cell* **120**, 15–20.

Lund, E., Guttinger, S., Calado, A., Dahlberg, J. E., and Kutay, U. (2004). Nuclear export of microRNA precursors. *Science* **303**, 95–98.

Olsen, P. H., and Ambros, V. (1999). The *lin-4* regulatory RNA controls developmental timing in *Caenorhabditis elegans* by blocking *lin-14* protein synthesis after the initiation of translation. *Dev. Biol.* **216**, 671–680.

Paddison, P. J., Caudy, A. A., Bernstein, E., Hannon, G. J., and Conklin, D. S. (2002). Short hairpin RNAs (shRNAs) induce sequence-specific silencing in mammalian cells. *Genes Dev.* **16**, 948–958.

Pfeffer, S., Zavolan, M., Grasser, F. A., Chien, M., Russo, J. J., Ju, J., John, B., Enright, A. J., Marks, D., Sander, C., and Tuschl, T. (2004). Identification of virus-encoded micro-RNAs. *Science* **304**, 734–736.

Pfeffer, S., Sewer, A., Lagos-Quintana, M., Sheridan, R., Sander, C., Grasser, F. A., van Dyk, L. F., Ho, C. K., Shuman, S., Chien, M., Russo, J. J., Ju, J., *et al.* (2005). Identification of microRNAs of the herpesvirus family. *Nat. Methods* **2,** 269–276.

Rodriguez, A., Griffiths-Jones, S., Ashurst, J. L., and Bradley, A. (2004). Identification of mammalian microRNA host genes and transcription units. *Genome Res.* **14,** 1902–1910.

Samols, M. A., Hu, J., Skalsky, R. L., and Renne, R. (2005). Cloning and identification of a microRNA cluster within the latency-associated region of Kaposi's sarcoma-associated herpesvirus. *J. Virol.* **79,** 9301–9305.

Sano, M., Kato, Y., and Taira, K. (2006). Sequence-specific interference by small RNAs derived from adenovirus VAI RNA. *FEBS Lett.* **580,** 1553–1564.

Schwarz, D. S., Hutvagner, G., Du, T., Xu, Z., Aronin, N., and Zamore, P. D. (2003). Asymmetry in the assembly of the RNAi enzyme complex. *Cell* **115,** 199–208.

Silva, J. M., Li, M. Z., Chang, K., Ge, W., Golding, M. C., Rickles, R. J., Siolas, D., Hu, G., Paddison, P. J., Schlabach, M. R., Sheth, N., Bradshaw, J., *et al.* (2005). Second-generation shRNA libraries covering the mouse and human genomes. *Nat. Genet.* **37,** 1281–1288.

Stegmeier, F., Hu, G., Rickles, R. J., Hannon, G. J., and Elledge, S. J. (2005). A lentiviral microRNA-based system for single-copy polymerase II–regulated RNA interference in mammalian cells. *Proc. Natl. Acad. Sci. USA* **102,** 13212–13217.

Sullivan, C. S., Grundhoff, A. T., Tevethia, S., Pipas, J. M., and Ganem, D. (2005). SV40-encoded microRNAs regulate viral gene expression and reduce susceptibility to cytotoxic T cells. *Nature* **435,** 682–686.

Yekta, S., Shih, I. H., and Bartel, D. P. (2004). MicroRNA-directed cleavage of HOXB8 mRNA. *Science* **304,** 594–596.

Yi, R., Qin, Y., Macara, I. G., and Cullen, B. R. (2003). Exportin-5 mediates the nuclear export of pre-microRNAs and short hairpin RNAs. *Genes Dev.* **17,** 3011–3016.

Zeng, Y., Cai, X., and Cullen, B. R. (2005). Use of RNA polymerase II to transcribe artificial microRNAs. *Methods Enzymol.* **392,** 371–380.

Zeng, Y., Wagner, E. J., and Cullen, B. R. (2002). Both natural and designed microRNAs can inhibit the expression of cognate mRNAs when expressed in human cells. *Mol. Cell* **9,** 1327–1333.

Zeng, Y., Yi, R., and Cullen, B. R. (2003). MicroRNAs and small interfering RNAs can inhibit mRNA expression by similar mechanisms. *Proc. Natl. Acad. Sci. USA* **100,** 9779–9784.

Author Index

Subject Index

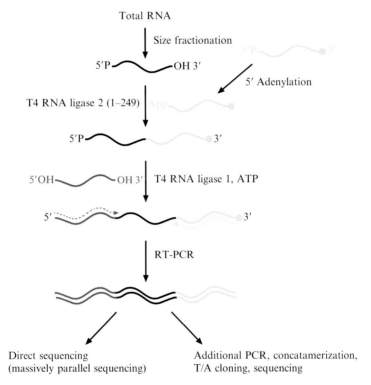

Sébastien Pfeffer, Figure 3.1 Schematic representation of the small RNA cloning protocol steps.

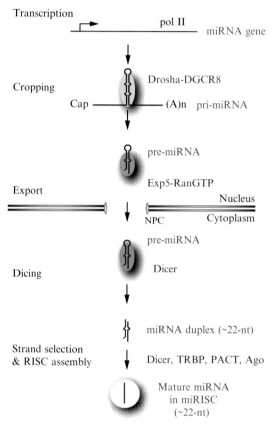

Yoontae Lee and V. Narry Kim, Figure 5.1 Current model for microRNA maturation.

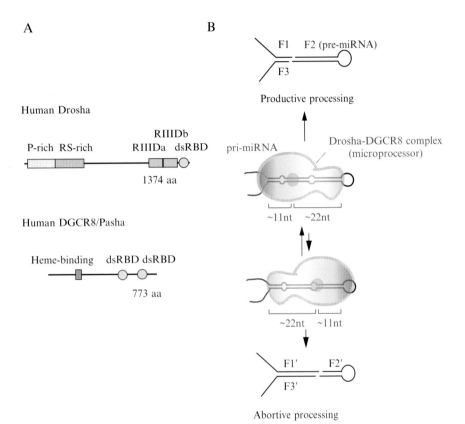

Yoontae Lee and V. Narry Kim, Figure 5.2 The microprocessor. (A) Schematic presentation of the domain structures of human Drosha and DGCR8. (B) The model for substrate recognition and cleavage site selection by the microprocessor.

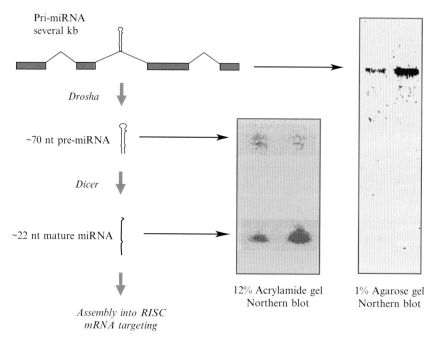

Pri-miRNA
several kb

Drosha

~70 nt pre-miRNA

Dicer

~22 nt mature miRNA

Assembly into RISC
mRNA targeting

12% Acrylamide gel
Northern blot

1% Agarose gel
Northern blot

J. Michael Thomson *et al.*, Figure 6.1   Biogenesis of miRNAs. The biogenesis steps for miRNAs are shown. After the mature miRNA species is produced, it is incorporated into the RNA–induced silencing complex (RISC) where it directs repression of targeted mRNAs (not shown). Northern blot detection of each intermediate is shown for illustrative purposes.

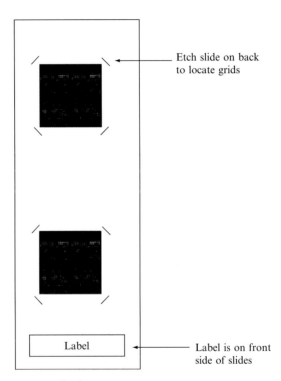

Etch slide on back
to locate grids

Label

Label is on front
side of slides

J. Michael Thomson *et al.*, **Figure 6.2**   Appearance of microRNA microarray slide.

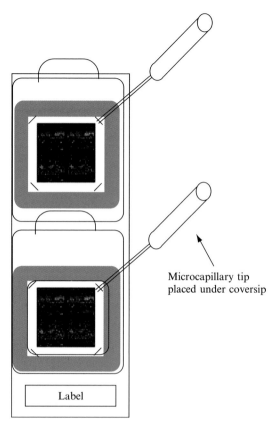

J. Michael Thomson *et al.*, Figure 6.3 Positioning of Frame–Seal and loading tip.

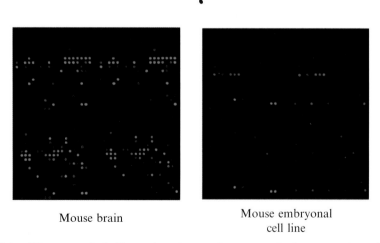

Mouse brain

Mouse embryonal
cell line

J. Michael Thomson *et al.*, Figure 6.4 Images from scanned microarrays. Scans from two microarray experiments are shown for illustrative purposes.

A   **2.5 to 5 ugms total RNA**

1) Primer annealing

5′   **miRNAs**   3′   2XBIOTIN

3′   R. Octamer P.   5′

2) RT ↓

3) RNA hydrolysis ↓

Reverse transcription with
biotin-labeled random
octamer primer

↓

Glass slide with 40 mer sense
oligo captured probe spotted
multiple times and in several places

↓

Streptavidin alexa 647

↓

Laser microarray scanner
measuring signal intensity
(miRNA abundance)

↓

Target RNA labeling

DNA–DNA hybridization

Staining

Signal detection

Figure 11.1   (*continued*)

B

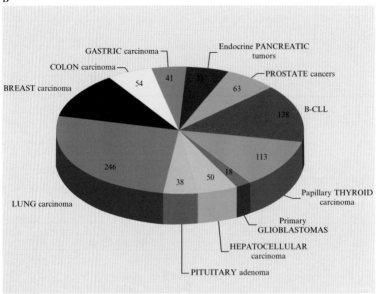

George Adrian Calin and Cario Maria Croce, Figure 11.1 Principles of microarray technology used for microRNA profiling. (A) The microarray-based miRNA profiling is presented as described in the majority of profiling studies on primary tumors, initially developed by Liu *et al.* (2004). Target labeling, hybridization, staining, and signal detection are the four main technical steps (presented on the right side). The different replicates of the spots on the glass slide (presented as different types of gray grids) represent different oligonucleotide sequences corresponding to sequences from the precursor miRNA or active miRNA molecule. The main advantage of the microarray-based miRNA profiling is the high standardization of the procedure (allowing the processing of tens of samples in parallel), making it relatively easy to be performed. Several technical aspects regarding this technology can be found elsewhere (Liu *et al.*, 2007). (B) Profiling of 921 hematological and solid tissue primary cancers (and corresponding normal tissues samples) performed by the microarray technology developed by Liu *et al.* (2004) as published to date (numbers are according to data compiled in Calin and Croce [2006a] and from new reports of Bottoni *et al.* [2005] and Weber *et al.* [2006]). The general consensus is that profiling allowed the definitions of signatures associated with diagnosis and prognosis. Modified with permission from *Nature Reviews Cancer* (Calin, G. A., and Croce, C. M. [2006]. MicroRNA signatures in human cancers. *Nat. Rev. Cancer* **6,**857–866.) Copyright (2006) MacMillan Magazines Ltd.